System of Systems Collaborative Formation

Systems Research Series — Vol. 1

System of Systems Collaborative Formation

Michael J. DiMario

Stevens Institute of Technology, USA

NEW JERSEY • LONDON • SINGAPORE • BEIJING • SHANGHAI • HONG KONG • TAIPEI • CHENNAI

Published by

World Scientific Publishing Co. Pte. Ltd.
5 Toh Tuck Link, Singapore 596224
USA office: 27 Warren Street, Suite 401-402, Hackensack, NJ 07601
UK office: 57 Shelton Street, Covent Garden, London WC2H 9HE

British Library Cataloguing-in-Publication Data
A catalogue record for this book is available from the British Library.

SYSTEM OF SYSTEMS COLLABORATIVE FORMATION
Systems Research Series — Vol. 1

ISBN-13 978-981-4313-88-9
ISBN-10 981-4313-88-2

Typeset by Stallion Press
Email: enquiries@stallionpress.com

Printed in Singapore by B & Jo Enterprise Pte Ltd

CONTENTS

Contents

LIST OF TABLES

LIST OF FIGURES

Appendix C

Appendix D

Appendix E

Appendix F

Appendix G

LIST OF SYMBOLS

u_i	Constituent System Utility
U	System of Systems Utility
Σ	Summation
$f(\cdot)$	Social Function
$\in$	Element of
$>$	Greater than
$\lambda_{\max}$	Largest Eigen Value
$\succ$	Preferred
E^3	System of Systems Environmental Cube (Systems, Capabilities, Time)
EU	Expected Utility

Chapter 1

INTRODUCTION

There is notably a growing interest in System of Systems (SoS) concepts and strategies. The performance optimization among groups of heterogeneous systems to realize a common goal has become the focus of various application areas including military, security, aerospace, and disaster management. There is particular interest in achieving synergy between these independent systems to enable the desired overall system performance. In the literature, researchers have begun to address the issue of coordination and interoperability in a SoS pointing to the emergence of the concept of system of systems engineering (SoSE). SoSE presents new challenges that are related to, but distinct from, systems engineering (SE) challenges. By understanding these differences, appropriate methods, tools, and standards, a SoSE approach may be crafted in an intelligent manner to architect and control seemingly amorphous systems.

1.1 Introduction and Overview

The engineering and control of emergent capabilities of large complex systems comprised of numerous seemingly unplanned contributing systems has been an unattainable goal of technologists and social infrastructure planners. Similarly described as an "invisible hand" in the published work of the Scottish philosopher Adam Smith in Book IV of the "Wealth of Nations" whereby economic processes, acting as individual agents to maximize their own well being, affect other processes without due intent (Smith, 1776). The invisible hand metaphor is intended to explain that actions have unintended consequences, are not controlled by a central command authority, and have an observable and patterned effect on the process and systems. The cooperative systems' architecture, design, and control are characterized by a set of interconnected systems with a common goal. The real-time cooperative control is not well understood let alone intentionally designed (Cloutier, DiMario, & Polzer, 2009; Tien, 2009). To possess the methodology to direct and control similar examples such as the U.S. health care system, world political systems, the effects of an economic recession, and a global shipping system is a study of complex systems.

Many complex systems are a SoS that comprise numerous constituent interdependent systems and have been described as an integration of complex metasystems — defined as a group of systems that have an interrelationship (Keating *et al.*, 2003). This system description supposes that the systems are autonomous and heterogeneous forming partnerships whereby their interoperability relationship produces capabilities or unintended consequences because of emergent behavior. The emergent behavior does not originate from any single individual constituent system nor deduced by properties of the collective constituent systems. The interoperability relationships of the constituent systems create new behaviors of the holistic system.

General Systems Theory (GST) introduces the holistic concept of systems as a science of wholeness whereby there are general systems laws, which apply to any system of a certain type independent of its properties, classified as summative and constitutive. The summative system owes its capabilities to the summation of the characteristics and sub capabilities of its elements. The properties of the elements are universally the same in all environments and have no relationships. The constitutive system has capabilities that are greater than the summation of its elements because the effects are dependent upon the context of their properties and their relationships (Bertalanffy, 1969). Kenneth Boulding a general systems theorist, described this concept as a SoS which may perform the function of a *gestalt*, which is a pattern so unified as a whole that its properties cannot be derived from its parts (Boulding, 1956).

This thesis discusses a constitutive SoS framework in regard to new behaviors that emerge as a result of a mechanism of collaboration that is architected to affect a holistic capability. The management and design of SoS architectures are of great interest as individual systems become ever more interoperable with other systems (U.S. Department of Defense, 2008). The ability to design and control a SoS architecture to elicit capabilities not germane to any one individual system becomes paramount (DiMario, Cloutier, & Verma, 2008). SoS architecture management and development of such structures, as well as the discipline of SoS, is of great interest with many avenues to explore as well as its foundational engineering science discussed further in Appendix C (DeLaurentis *et al.*, 2007).

1.2 Problem Statement and Research Hypothesis

A description of an approach of a mechanism framework whereby the paradox of autonomous and yet cooperative systems may be architected to elicit SoS capabilities is lacking. A methodology and framework to architect systems that may influence other systems in collaboration is absent in the systems literature as well as literature in support of systems that naturally interoperate but were not presciently designed to do so. The research hypothesis is a system of systems (an SoS), as distinct from a system of parts, is a system comprised of pre-existing autonomous systems that choose to belong to the dynamically forming SoS, and interoperate together in spite of their evident diversity, is a result of changes in the autonomous systems' environment or their own utility resulting in capabilities of the SoS not presciently designed.

1.3 Research Objectives

The research objectives are to describe the influence of SoS characteristics that demarcate a class of systems defined as constituent — design-time SoS, and run-time SoS. This includes a description of how constituent systems may cooperate through the characteristic of belonging creating value of the SoS as a constitutive mechanism. This is done via design rules or a social function that balances the SoS characteristics of autonomy and belonging. SoS run-time and design-time are uniquely assembled by interoperability at various levels and define system's belonging.

1.4 Uniqueness of this Research

This research addresses the autonomy of systems, managed independently, to make independent choices of collaboration. However, the SoS cannot exist physically if not for constituent systems agreeing to be "integrated" and having a sense of belonging. For an SoS, the hypothesis of this thesis is a mechanism that allows systems to be integrated at a holistic level that are either designed, referred to as design-time, or occur naturally via self-organization and referred to as run-time. Examples are:

- Al-Qaeda: Run-time SoS;
- Health care: Run-time SoS;
- USN Battle Group: Design-time and run-time SoS;
- National Transportation: Run-time SoS;

- Coast Guard Deepwater Program: Design-time SoS at end of the program or T_0; Expected run-time at later epochs or T_1.

The constitutive mechanism defines the levels of belonging as a design for influence approach defines the design rules or rules naturally evolve as an "invisible hand" or amorphous via self-organization.

1.5 Dissertation Organization and Structure

This chapter provides an introduction and context for this research. Chapter two highlights the relevant literature of SoS and an esemplastic, a union of ideas giving way to new concepts, approach to the problems presented. Chapter three discusses the research approach. Chapter four discusses the results of the research and findings. Chapter five discusses the validation of the research as a case study approach of low-constraint methods and the case study data. Chapter six highlights the research conclusions and next logical research steps. The remainder of the dissertation consists of appendices of published or to be published papers, additional research supporting material, and dissertation references.

Chapter 2

LITERATURE REVIEW AND RESEARCH BOUNDARY

There is sparse literature concerning the formation of a SoS. The systems engineering discipline of SoS is in its formative phase (Gorod, Sauser, & Boardman, 2008). The approach in the literature review and in this research is an esemplastic approach that includes systems thinking, systems engineering concepts, social networks, mechanisms of economics, and decision theory:

- Systems Thinking — provide a foundation for SoS as it has provided the same for systems engineering (Arquilla & Ronfeldt, 2001; Bertalanffy, 1969; Boardman & Sauser, 2008; Boulding, 1956; Checkland, 1993, 2000; Holland, 1998; Herbert A. Simon, 1962);
- SoS — the literature is rich in SoS definitions, perspectives, abstractions, and general notions of what a SoS is or is not. Other than definitions, the literature is largely void concerning SoSE, SoS management, developing an SoS, or the understanding of naturally occurring (i.e. non-developed or unplanned) SoS structures. The uniform agreement in the literature is that a SoS is comprised of cooperative constituent systems (Alberts, Garstka, & Stein, 2000; Boardman, 1995; Krygiel, 1999; Maier, 1996; Morganwalp, 2002; Sage & Cuppan, 2001);
- Systems Cooperation — the literature does not address how constituent systems interoperate but assumes interoperation is by typical approaches of preplanned syntactic connectivity. This assumption accounts for man-made systems, but does not account for naturally occurring SoS that have evolved over long periods of time such as the National Transportation System, U.S. health care, and the Al-Qaeda terrorist organization. Systems cooperation may illicit itself in any manner by which systems may interoperate via direct communication through communication protocols or indirectly through forms of stigmergy. The reason why systems cooperate, especially in real-time is largely absent in the SoS

literature but may be found as an esemplastic union subject of economics and game theory known as mechanisms (Axelrod, 2006; Kim & Roush, 1987; Parunak, 1997; H. A. Simon, 1955; Herbert A. Simon, 1956; Stirling, 2003, 2005; Stirling & Frost, 2005);

- Mechanisms — For a SoS to be architected and managed, mechanism design found in economics, political science, and game theory is the subject of designing rules of a system to achieve a specific outcome, even when the system may be autonomous and of self-interest (Myerson, 1984). For game theory, the purpose of mechanism design is to architect the rules of the environment or the rules the agents are to play for maximization of the collective good of the agents (Russell & Norvig, 2003). A mechanism that allows for a constitutive SoS provides a holistic description and a basis of decision to how and why systems as a collective choose to cooperate and define the nature of the SoS characteristics. A fundamental premise adopted from game theory is that not all constituent systems behave the same and are autonomous and heterogeneous. With this understanding, the environment by which systems cooperate, as well as the systems themselves, is designed and managed to achieve a common goal even though the holistic behavior appears to be amorphous. The ability for an autonomous constituent system to decide to cooperate via a mechanism is investigated in the interdisciplinary subject of decision theory (Goodrich, Stirling, & Frost, 1998; Hurwicz & Reiter, 2006; Myerson, 1984; J. Nash, 1951; J. F. Nash, 1950; Von Neumann & Morgenstern, 1953);
- Decision Theory — A fundamental basis of decision theory is a system that is rational if and only if it chooses actions that yield the highest utility. A system's belief state is a representation of the probabilities of all possible actual states of the environment. The choice of an action of a system state is determined by the comparison of utilities when there is risk or uncertainty. For rational systems, the choice of an action is with the maximum expected utility (*EU*). The holistic collective benefit through the individual system's participation and cooperation is reached when each system adopts a solution that maximizes its own utility function (u_i) or self-interest. The decision to maximize the system's EU is in response to its perception of the environment. If probabilities are unknown, a number of methods have been proposed such as choose any systems of weights or maximize the minimum results (Figueira, Salvatore, & Ehrgott, 2005; Keeney, 1992; Keeney & Raiffa, 1993; Saaty, 2005; Saaty & Vargas, 2001).

2.1 Systems Thinking

The holistic concept of systems is not new as it was introduced in GST as a science of wholeness whereby there are general systems laws that apply to any system of a certain type independent of its properties. The elements of a system are classified into two types as *summative* and *constitutive.* The summative system owes its capabilities to the summation of the characteristics and sub capabilities of its elements. The properties of the elements are universally the same in all environments and have no relationships. The constitutive system has capabilities that are greater than the summation of its elements because the effects are dependent upon the context of their properties and their relationships (Bertalanffy, 1969). Boulding described GST as an optimum degree of generality that is not reached by the particular sciences and is somewhere between a theory without content and too much content where similarities among seemingly disparate systems are obscured. He further described this concept of a spectrum of theories as a SoS which may perform the function of a *gestalt*, which is a pattern so unified as a whole that its properties cannot be derived from its parts (Boulding, 1956).

INCOSE defines a system as an integrated set of elements that accomplish a defined objective. These elements include products (hardware, software, and firmware), processes, people, information, techniques, facilities, services, and other support elements (INCOSE, 2007). This definition accentuates the components or groupings of objects to perform a task or an activity. In concert with GST, systems are defined as theoretical objects with regard to their activities of transformation in which they are involved. Systems transform some states of their environment into other states in their environment. They respond to variation in their environment by producing certain effects upon that environment. The act of transformation defines them as systems and focuses not on the components but on the activities of the whole. In this manner, systems are constructs defined with regard to their activity and not their components.

The quality of interactions or interoperation of systems is based on a degree of coordination, cooperation, and collaboration leading to a consensus. The spectrum of the quality of interactions or reasons for coexistence reflects the attractiveness of constituent systems to belong to a greater whole and thus identifying principles of togetherness (Boardman & Sauser, 2008). Patterns of interactions remain stable as long as they do not produce an output that is persistently "out of balance" with regard to its inputs. Logic of system interdependence is bounded by mutual limitations of a range of possible qualitative structures that are compatible with interactions of each participating constitute system.

Coordination — defined as two or more disparate systems working together for a goal that must harmonize in a common effort. The behaviors of the entities must align spatially and temporally and may be represented in three different perspectives as the coordination of action, coordination of understanding, and coordination of goals (Arrow, McGrath, & Berdahl, 2000). The members of the group agree on shared meanings and norms about who should do what, when, and how.

Cooperation — defined as two or more disparate entities to work together toward a common goal of mutual benefit. The entities work together to achieve global properties. Cooperation may be coerced, voluntary, or unintentional. Self-interest of the entity may not favor cooperation as shown in the Prisoner's Dilemma (Axelrod, 2006) by acting to sacrifice their individual interests for the sake of the common good, or to further their own selfish individual interests at the expense of the other. This was also shown to be the opposite situation when cooperation is recursive in the Iterated Prisoner's Dilemma (Mitchell, 1998). However, the relationship of two entities is binary until there is a third. A third entity introduces a network whose systemic traits are more than the sum of each of the individuals attributes. A triad is the basis of social interaction (Degenne & Forse, 1999) and is considered in this framework of semantic interoperability of systems.

Collaboration — defined as two or more disparate entities share their individual information, awareness, and understanding to shared and common information, awareness and understanding. The nature of collaboration, speed of collaboration, and likelihood of successful collaboration is dependent on this ability in which to share. Allocation of decision rights and information distribution and interactions are permitted and granted by individuals. This is also dependent on Community of Interest (COI) composed of all actors who care about and can influence decisions (Cloutier *et al.*, 2009). Maximum collaboration attributes are inclusive of all entities assuming all entities have equal and full access to all other entities. This is performed via unconstrained connectivity by which it is participatory and continuous. Information exchanged is complete including actions, decisions, knowledge, and understanding. The disparate entities will sacrifice their individual objectives for common group objectives.

Consensus — even though coordination, cooperation, and collaboration are degrees of agreement, the same degrees apply in converse as disagreement or competition. Consent requires proper framing of goals and within the realm of the possible as evidence in the role of Admiral Nelson in the Battle of Trafalgar (Alberts & Hayes, 2003) through unity of purpose via constructive interdependence. Although there is a state of governance in the Battle of Trafalgar, in a SoS consensus is the idea that entities will form strong agreement on what needs to be done and how to do it that leads to at least a temporary union or coalition and is societal in nature. It is also recognized that this reduces diversity and actually may reduce organizational learning (Flood, 1999) as independence may be lost. However, the disparate entities seeking accommodation find common ground while preserving unique diversity and autonomy or finding common ground while preserving differences.

2.2 System of Systems

The general concept of SoS is most notably associated with the U.S. Department of Defense (DoD) for applications of Network Centric Warfare (NCW) (Cloutier *et al.*, 2009). The SoS term is used to describe capabilities resulting from the ubiquitous connectivity of systems and is generally associated with large-scale inter-disciplinary objectives with multiple, heterogeneous, and distributed systems that are at various levels and domains. An early adopter of the term, Admiral William Owens, Vice Chairman of the Joint Chiefs of Staff in 1996, used the term in his description of the need to improve intelligence, surveillance, and reconnaissance (ISR) to reduce the "fog of war" (Owens, 1996). Since the creation of the DoD SoS associated concepts, other terms and concepts have emerged such as family of systems (FoS) (U.S. Joint Chiefs of Staff, 2005), and federation of systems (F-SoS) (Krygiel, 1999) as well as greater than forty variant definitions (Boardman & Sauser, 2006) and are inconsistent as well (Saunders *et al.*, 2005). The concept may be associated with DoD applications, but SoS is not foreign to non-DoD domains (Gorod *et al.*, 2008). SoS naturally exists where there are systems communicating with other systems such as in a large transportation system (DeLaurentis, Lewe, & Schrage, 2003), space exploration (Aldridge, 2004), health care (Reid *et al.*, 2005), vital infrastructures, and climate change.

System of systems comprises numerous constituent interdependent systems and has been described as an integration of complex metasystems as well as a potentially new field of systems engineering or at least an extension to traditional systems engineering (TSE) to develop and manage interconnected systems (Keating *et al.*, 2003). These systems are autonomous and heterogeneous forming partnerships whereby their interoperability relationship produces capabilities or unintended consequences that do not originate from anyone individual constituent system. The interoperability relationships of the constituent systems create new behaviors of the holistic system. The picture that emerges is one of a holistic versus a constituent perspective. Linking systems to form SoS through interoperability and synergism allows for emergent behaviors providing valued capabilities as well as problems that are not resident in any of the constituent systems that comprise a SoS. The interoperability of multiple disparate systems resulting in effects that are based on the whole system leads to a holistic systems approach.

As there are numerous definitions of a SoS, there are numerous definitions of a system. The INCOSE organization and the ISO/IEC 15288 standard define a system as "a combination of interacting elements organized to achieve one or more stated purposes." These elements include products (hardware, software, and firmware), processes, people, information, techniques, facilities, services, and other support elements (INCOSE, 2007; ISO, 2002). Buede defines a system as "a set of components (subsystems, segments) acting together to achieve a set of common objectives via the accomplishment of a set of tasks" (Buede, 2000). For the purposes of this thesis, we will consider a system as something that has the ability to perform a set of tasks to satisfy a mission or objective.

The DoD has defined a SoS "as a set or arrangement of systems that results when independent and useful systems are integrated into a larger system that delivers unique capabilities" (DoD, 2006). INCOSE defines a SoS as "a system-of-interest whose system elements are themselves systems; typically these entail large scale inter-disciplinary problems with multiple, heterogeneous, distributed systems" (INCOSE, 2007). This definition and the various other definitions of systems and SoS do not provide a detailed taxonomy to provide differentiation of a system versus a SoS resulting in a dual perspective of a system-of-interest is my element versus your system. This perception extends to your system is my SoS resulting in sufficient discussion on differentiating elements, systems, and SoS (Boardman, 1995; DeLaurentis & Crossley, 2005; Koestler, 1967; Herbert A. Simon, 1962, 1996).

Arthur Koestler managed this perceptual challenge with the introduction of *holons*, which are nodes on a hierarchic tree that behave partly as systems, or as elements, depending on how the observer perceives them (Koestler, 1967). In this model, there are intermediary forms of a series of levels in an ascending order of complexity. These levels are considered sub-systems that have characteristics of the system as well as its elements. This leads to the Janus Principle, so named after a two faced Roman god that guarded the doors of the Pantheon and could see within the Pantheon as well as outside, whereby within the concept of hierarchic order, members of the hierarchy have two faces looking in two directions. One face is looking up the hierarchy, and the other looking down the hierarchy. The member as a system is looking downward in descending order to less complex levels while the member as an element is looking upward in ascending order to more complex levels and is the dependent part of the hierarchy without an understanding how the lower levels are organized to form the whole (Driscoll, 2008).

Peter Checkland adds further discussion to Boulding's and von Bertalanffy's thesis that hierarchy provides both an account of the relationships between different levels and an account of how observed hierarchies are formed giving rise to holistic behavior based on interdependence through a means of interoperability. Checkland offers the following definitions (Checkland, 1993):

- **Connectivity:** In the formal system model, the property that enables effects to be transmitted through the system. The connectivity may have a physical embodiment (as in an order processing system) or may be a flow of energy (verbal) information or influence.
- **Hierarchy:** The principle according to which entities meaningfully treated as wholes are built up of smaller entities which are themselves wholes... and so on. In a hierarchy, emergent properties denote the levels.
- **Emergence:** The principle that whole entities exhibit properties that is meaningful only when attributed to the whole, not to its parts.

So what is an element versus a system versus a SoS? The proper system's spectrum inter-relationship taxonomy of elements, components, subsystems, systems, and the various forms of SoS, such as FoS, F-SoS, as a common frame of reference does not suitably exist. The Janus effect will be present in the systems engineering community until a suitable lexicon is devised. However, SoS through the interdependence of traditional systems have characteristics of autonomy, belonging, connectivity, diversity, and emergence that a non-interdependent traditional system does not possess (Boardman & Sauser, 2006).

The system's spectrum consists of components on one end through traditional systems to SoS comprised of constituent systems on the other furthest end. The corresponding attributes range from one of central control for a monolithic or unitary system to a mixture of attributes for a decentralized system to a set of attributes for a SoS which is the dichotomy of the systems and elements of autonomy and dependence. A primary assumption of a SoS is that they are open systems and are best understood in the context of their environments. It is proposed that SoS have unique attributes unlike traditional ordinary or single systems (Boardman & Sauser, 2006; DeLaurentis & Crossley, 2005; Keating *et al.*, 2003; Maier, 1996; Sage & Cuppan, 2001; Saunders *et al.*, 2005). A set of attributes, autonomy, belonging, connectivity, diversity, and emergence, are used for reference are shown in Table 1 as they relate to a system and a SoS (Boardman & Sauser, 2006). A SoS is comprised of autonomous constituent systems fulfilling an objective as independent systems but are interdependent via interoperability to fulfill holistic objectives of a SoS. The constituent systems must perform an interoperation to fulfill a SoS capability in a collective manner whereby the capability does not exist on any of the constituent systems.

SoS behavior and capabilities are that of the whole and not properties of the constituent systems or that can be deduced by properties of the elements. The emergent behavior is a product of the interactions and not the sum of the actions of the constituent elements. Behavior arises from the organization of the elements, vis-à-vis the constituent systems without having to identify the properties of the individual elements. For example, a computer is an organization of elementary functional components whereby the arrangement of the components describes the behaviour of the whole system (Herbert A. Simon, 1996).

2.3 Systems Cooperation

2.3.1 *Connectivity and Interoperability*

Connectivity is the interoperability of the constituent systems expressing their requirements and needs. Connectivity may comprise any means necessary to link and to send information between constituent systems of a SoS. The internal machinery of the receiving system makes use of

Table 1. Attribute Differentiation of a System and System of Systems.

Attribute	System	System of Systems
Autonomy	Autonomy is ceded by parts in order to grant autonomy to the system.	Autonomy is exercised by constituent systems in order to fulfill the purpose of the SoS.
Belonging	Parts are akin to family members; They did not choose themselves but came from parents. Belonging of parts is in their nature.	Constituent systems choose to belong on a cost/benefits basis; also in order to cause greater fulfillment of their own purposes, and because of belief in the SoS supra purpose.
Connectivity	Prescient design, along with parts, with high connectivity hidden in elements, and minimum connectivity among major subsystems.	Dynamically supplied by constituent systems with every possibility of myriad connections between constituent systems, possibly via a net-centric architecture, to enhance SoS capability.
Diversity	Managed i.e. reduced or minimized by modular hierarchy; parts' diversity encapsulated to create a known discrete module whose nature is to project simplicity into the next level of the hierarchy.	Increased diversity in SoS capability achieved by released autonomy, committed belonging, and open connectivity.
Emergence	Foreseen, both good and bad behavior, and designed in or tested out as appropriate.	Enhanced by deliberately not being foreseen, though its crucial importance is, and by creating an emergence capability climate, that will support early detection and elimination of bad behaviors.

the information for the benefit or detriment of the SoS. In order to be persistent and common, connectivity may range in composition from chemical communication of bacteria (Bassler & Losick, 2006) to networks of computers and society. The connectivity characteristic is an interoperability enabler and is discussed extensively in Appendix D (DiMario, 2006). A definition used in this context is: "The ability of a collection of communicating entities to (a) share specified information and (b) operate on that information according to a shared operational semantics in order to achieve a specified purpose in a given context" (Carney, Fisher, Morris, & Place, 2005).

Systems interoperability or interoperation assumes that systems have a relationship and that a system provides a service to another. In other words, in a SoS context there are cooperative interactions among loosely cooperative systems to adaptively fulfill a system-wide purpose or are encapsulated within an enterprise architectural framework (Morganwalp, 2002). Syntactic communication is the initial step of interoperability whereby a connection is established but semantic communication is necessary for the communicating systems to know what to do with the information that is communicated creating value (van Diggelen, 2007). SoS interoperability (SoSI) is fundamental to the SoS complex to provide value that is greater than the sum of its individual participants. The systems are "SoS enabled" whereby their interoperability is based on a convergence protocol or the interfaces are ubiquitous. The convergence protocol is omnipresent and will be used when a system adds value or dynamically belongs to the SoS collective. The key point is for a group of system participants through a readily available interoperability schema to elicit value greater than their own without syntactic change.

SoS becomes realizable through the weaving of constituent systems via a connection. The connections may be dynamic and unpredictable that the SoS is said to be unbounded because the definition and capabilities of the SoS is indeterministic. Bounded systems typically rely on a hierarchical structure, central control, central data, and a high degree of trust. These properties are typically that of an independent system. Unbounded systems employ components never imagined to be used. However, one may consider that even though a SoS's internal connectivity may be dynamic and its future connectivity is unknown — for a point in time, there are known connections and thus the SoS is bounded. An SoSI taxonomy is suggested to further differentiate systems and a SoS (Majumdar, 2007):

Level 0: No connection.
Level 1: Contribute — situational awareness primarily for safety or information exchange.
Level 2: Coordinate — determination of how and when to share resources.
Level 3: Cooperate — determination and plan of how to accomplish related tasks for a common goal.
Level 4: Collaborate — determination and work to be done together to accomplish a task.

Table 2. System Characteristics of Integration vs. Interoperation.

	System Integration	**System Interoperation**
Process	Composing non-autonomous constituents	Composing autonomous constituents
Composition	Predefined	Loosely predefined to not defined
Structure	Hierarchical	Flat; Distributed
Control	Central	Distributed
Boundary	Fixed	Dynamic
System Coupling	Tight	Loose
Centricity	Node	System
Evolution	Resistant	Influence

An understanding of systems integration versus interoperation becomes a critical differentiator between constituent interdependent systems of a SoS and independent systems shown in Table 2. The integration and interoperation characteristics define basic system and SoS definitions. The critical aspect to recognize is that *integration* and *interoperation* are functions in this context. It is a function describing a process by which a system's parts (being subsystems or components) versus a SoS's parts (being autonomous constituent systems) are differentiated. A builder of large systems has an implementation choice of whether to place constituent system boxes into a single box and to meld them in such a way that constituent autonomy is lost or to build in such a fashion that autonomy is maintained and capabilities are developed via collaborative interoperation (Caffall, 2005).

2.3.2 *Belonging and Collaboration*

Interoperability is dependent on the characteristics of the constituent systems as well as the nature and interaction of other characteristics of other systems. In SoS, thus follows that the state of the SoS as a system is determined by its history of interactions that have occurred thus far. Systems follow their own interest and the resultant system formed via interoperability is based on the individual objectives of the constituent systems and the behavior of the whole can be only estimated at best via deduction. What is important for SoS is to discover what is necessary for cooperation to emerge, appropriate actions can be taken to foster development of cooperation in a specific setting. Cooperation is based on an actor's interest to cooperate. There are certain cases in which cooperation is not completely based upon a concern for others or upon the welfare of the group as a whole as demonstrated in the popular iterated prisoner's dilemma (Axelrod, 2006). This form of the prisoner's dilemma is a two-player game whereby players try to maximize their performance or fitness by cooperating with the other player or defecting. The payoff received is dependent on the actions of the other player. Axelrod set up experiments in which strategies were evolved using as their fitness the mean score attained against a set of eight human strategies. He illustrated that the system evolved the best strategy.

The analogy to demonstrate cooperation and a sense of belonging is taken from biological organisms as mutualism that is an association, as symbiosis or parasitism, of two organisms. The cooperation of organisms for a mutual benefit is symbiosis and a negative effect is parasitism. Parasitism results in the death of both organisms whereby symbiosis results in coadaptation and coevolution. Symbiosis can also result in endosymbiosis whereby two species become so interdependent that they become inextricably linked. This is akin to different parts of a problem is so interdependent that they are not amendable to division.

The symbiotic analogy extends to systems as organisms. A system is said to be endosymbiotic when it cannot be separated or decomposed and yet fulfill its intended purpose and loses its autonomy. A system is symbiotic whereby it cooperates and shares its capabilities and resources with another system to fulfill a mutually agreed upon need. This is shown as an extraction of Table 2 and shown in Table 3 whereby a system that is integrated is interdependent and thus is endosymbiotic. A system that is interoperable is symbiotic and maintains its autonomy.

2.3.3 *System of Systems Interoperability and Satisficing*

SoS is realized through the weaving of constituent systems via multiple connections. The connections are direct as logical or physical paths and

Table 3. Systems Cooperation Characteristics.

	Mutualism	Autonomy
System Integration	Endosymbiosis	No autonomy of endosymbiotic components or systems. Autonomy is displayed as the union or summation of the component systems.
System Interoperation	Symbiosis	Autonomy of symbiotic systems

indirect based on effects of influence such as how politicians are motivated by their constituency (Driscoll, 2008). The connections may be dynamic and unpredictable in such a manner that the SoS is said to be unbounded and adaptive because the definition and capabilities of the SoS are indeterministic (De Wolf, 2007). It must be noted that there is a difference between adaptive versus dynamic systems. For the case of an aircraft, the autopilot system is dynamic as it responds to variations via a set of rules. Adaptive is associated with the actual human pilots who respond to changing conditions with unknown rules. Adaptive systems are able to modify their output in such a manner that responds to unanticipated changes in the environment.

Systems are bounded as they rely on a hierarchical structure, central control, central data, and a high degree of trust. These properties are typically that of an independent and stable system optimized for improved performance. However, there is a paradox for optimized systems in a SoS context whereby too much efficiency impairs the ability to respond as the constituent systems lose their dynamic or adaptive capabilities. Optimized systems are inflexible, as their interfaces to other systems have fixed boundaries and tight coupling represented by a summative system versus dynamic boundaries and loose coupling represented by a constitutive system. Cooperative mechanisms of a constitutive SoS are likely to be satisficing, defined as a solution with acceptable intrinsic cost trade-off (West, Kobza, & Goerger, 2008).

The satisficing process, first suggested by Herbert Simon (H. A. Simon, 1955; Herbert A. Simon, 1956), of a holistic system includes the management of the constituent systems to achieve a favorable result for the user (Gorod, DiMario, Sauser, & Boardman, 2009). The process seeks an adequate versus an optimized solution because it is not possible to optimize for an individual system and the whole system simultaneously. It is difficult to define what is best for the group in terms of individual preferences (Stirling & Frost, 2005). Satisficing allows for degrees of fulfillment, whereas optimization is absolute. Simply stated, optimization in a SoS means — what is best for me is best for you which does not work well with other systems as there is no negotiation and lacks a trade-space. Satisficing in a SoS means — what is good for me is good for you which has room for negotiation and a trade-space.

The execution of cooperation in a satisficing process assumes that systems make rational choices and is the normative model. The decision-maker will estimate utilities of a choice, calculate the value, and choose the option with the largest value. Rational choice assumes decision makers have all available information that will influence a decision as well as the rewards. Simon (1956) proposed a bounded rationality whereby obtaining maximum utility as well as an optimized decision is not possible due to insufficient information and an inability to compare and weigh information against available options.

For the case of a SoS, the holistic system does not exist nor can be possibly satisficing or optimized if it were not for their interconnectivity. SoS characteristic connectivity is the interoperability of the constituent systems expressing their requirements and needs. Connectivity may comprise any means necessary, either directly or indirectly, to exchange information between constituent systems of a SoS. The internal machinery of the receiving system makes use of the information for the benefit or detriment of the SoS as an interdependent system.

2.3.4 *The Collaborative Formation of System of Systems*

A distinction adopted from (Carney, Fisher, Morris *et al.*, 2005) is how a SoS is formed by design-time or run-time processes. Design-time or preplanned SoS is when all components are determined and designed to cooperate as in the case of the U.S. Coast Guard Deepwater Program. This large program consists of a myriad of different classes of boats, ships and other assets designed to interoperate (Acquisition Solutions, 2001; Arnberg, 2005). As a function of time, each of the constituent systems will most likely evolve independently due to individual system managerial processes and system life cycles changing the formation of the SoS from design-time to run-time. From the example of a worldwide shipper, the shipper designs the SoS or grouping of services to get a package to the receiver and must respond to changes that effect the constituent systems to be employed.

Run-time SoS is composeable or formational in real-time whereby a new system is formed from constituents that are (a) compatible but were not designed to work together, (b) are incompatible and communicate in some manner, or (c) cooperate due to a specific need or out of self-interest. An example is when a warfighter "duck tapes" two systems, not designed to work together, to solve an immediate problem or provide an immediate service. The newly created system is the result of an interoperation of the two constituent systems by any means required by the warfighter, whether it is by compatible interfaces or data transfer via a memory stick. The newly constructed holistic system delivers a new capability that is not resident with its constituents. The new capability, however, may only be viable in this particular warfighter's environment as the required service is environment and user specific (Krygiel, 1999).

In another example, the U.S. health care system is amorphous whereby the large constituent systems such as hospitals, pharmaceutical companies, insurance companies, and physician's office networks are largely independent but are interdependent from the perspective of the user of the health care system (Institute of Medicine, 2001). The health care system provides value to the user because the interaction of the independent systems provides capabilities. An individual system or organization such as a hospital cannot offer total value if it were not for the pharmaceutical and insurance companies. The interdependence and constituent systems' cooperation is via a set of rules and processes that have evolved or created by government policy to accommodate their own self-interest as well as the health care user. In this manner, the individual systems are satisficing to provide services across a broad array of disparate health care systems enabling the health care SoS.

2.4 Mechanism

For a SoS to be architected and managed, mechanism design found in economics, political science, and game theory is the subject of designing rules of a system to achieve a specific outcome, even when the system may be autonomous and of self-interest (Myerson, 1984). For game theory, the purpose of mechanism design is to architect the rules of the environment or the rules the agents are to play for maximization of the collective good of the agents (Russell & Norvig, 2003). A mechanism that allows for a constitutive SoS provides a holistic description and a basis of decision to how and why systems as a collective choose to cooperate and define the nature of the SoS characteristics. A fundamental premise adopted from game theory is that not all constituent systems behave the same and are autonomous and heterogeneous. With this understanding, the environment by which systems cooperate, as well as the systems themselves, is designed and managed to achieve a common goal even though the holistic behavior appears to be amorphous (Gorod, DiMario, Sauser & Boardman, 2009).

The constitutive SoS collaborative mechanism may best be described as a global utility function (U). The SoS utility function exists via collective interdependent systems' architecture of composeability by design-time methods of systems engineering or via run-time methods of self-organization (DiMario *et al.*, 2008) and drawn from utility theory (Jensen & Nielsen, 2007; Russell & Norvig, 2003). It is an over simplification as rational systems weigh the advantages and disadvantages of cooperation and collaboration based on known and predicted facts which are subject to change. The SoS utility is a function of the value delivered to the users. To reach this holistic value, the individual participating systems weight their own self-interest gains against the costs of cooperation. As long as the acceptable benefits are greater than the costs, constituent systems will cooperate in a satisficing manner (Archibald, Hill, Johnson, & Stirling, 2006; Hurwicz, 1973). The maximization of a system's self-interest or utility function u_i underlies all economic theory and applications. Combined with mechanism design, there emerges a framework by which enterprises comprised of multiple disparate systems may be analyzed. It becomes the looking glass to reveal what is desirable and what is possible in the organization of systems (Ledyard, 1989).

2.5 Decision Theory

A fundamental basis of decision theory is a system that is rational if and only if it chooses actions that yield the highest utility. A system's belief state is a representation of the probabilities of all possible actual states of the environment. The choice of an action of a system state is determined by the comparison of utilities when there is risk or uncertainty. For rational systems, the choice of an action is with the maximum expected utility (*EU*). The holistic collective benefit through the individual system's participation and cooperation is reached when each system adopts a solution that maximizes its own utility function u_i or self-interest. The decision to maximize the system's EU is in response to its perception of the environment. If probabilities are unknown, a number of methods have been proposed such as choose any systems of weights or maximize the minimum results (Kim & Roush, 1987).

The maximization of the individual constituent system's utility function u_i is shown through the SoS characteristic of autonomy whereby each system does not know about the whole problem being solved, as in the case of the Janus Principle, just that its own utility is maximized. The constituent system maximizes its autonomy in fulfillment of its goals and participates in the collective in self-interest resulting in a global utility function (U). However, the simple summation of the individual constituent system's utility function does not necessarily contribute to the utility function of the whole. It results in a summative SoS and the constituents degrade the global utility because actions of each system effects the other such as the consumption of resources referred to as the tragedy of the commons (Russell & Norvig, 2003). For example, in the situation of a sporting event whereby thousands of enthusiastic fans simultaneously initiate cell phone calls to their friends to report on their team's latest score, the cell phone switching center quickly reaches over-capacity in handling the calls thus degrading cell phone usage for everyone. The cell phone users feel a right to make calls as they have paid for the service and do not relinquish usage. In this example, an individual has a wish to maximize their EU. However, either an individual system may maximize their own EU or that of the SoS, but not both; a situation of optimizing versus satisficing. If the SoS is to optimize its EU then the solution may not be acceptable to the individual constituent systems. However, a dominant strategy may exist whereby a strategy of cooperation yields a higher utility no matter what the other constituent systems do.

Chapter 3

RESEARCH APPROACH

The literature review of Chapter 2 provides an esemplastic description of the approach to the understanding of SoS theory and the supporting disciplines to provide a foundation to the understanding of how a SoS forms. The SoS formation is in regard to preplanned and architected systems and an SoS that forms via self-organization of systems. Both SoS methodologies include system interoperability since it is fundamentally assumed that a SoS requires a form of interoperation among its constituent systems to be viable. The research approach is shown in Figure 1 comprised of SoS design-time, SoS run-time, and SoS interoperability.

3.1 System of Systems Interoperability

SoS is realized through the weaving of constituent systems via multiple connections. The connections are direct as logical or physical paths and indirect based on effects of influence such as how politicians are motivated by their constituency (Driscoll, 2008). The connections may be dynamic and unpredictable in such a manner that the SoS is said to be unbounded and adaptive because the definition and capabilities of the SoS are indeterministic. There is a difference between adaptive versus dynamic systems. For the case of an aircraft, the autopilot system is dynamic as it responds to variations via a set of rules. Adaptive is associated with the actual human pilots who respond to changing conditions with unknown rules. Adaptive systems are able to modify their output in such a manner that responds to unanticipated changes in the environment.

The approach investigated for SoSI is of direct communications among constituent systems such as communications links (Morris, Levine, Meyers, Place, & Plakosh, 2004) or indirect such as stigmergy, the means by which insects and bacteria communicate (Bassler & Losick, 2006; Bonabeau

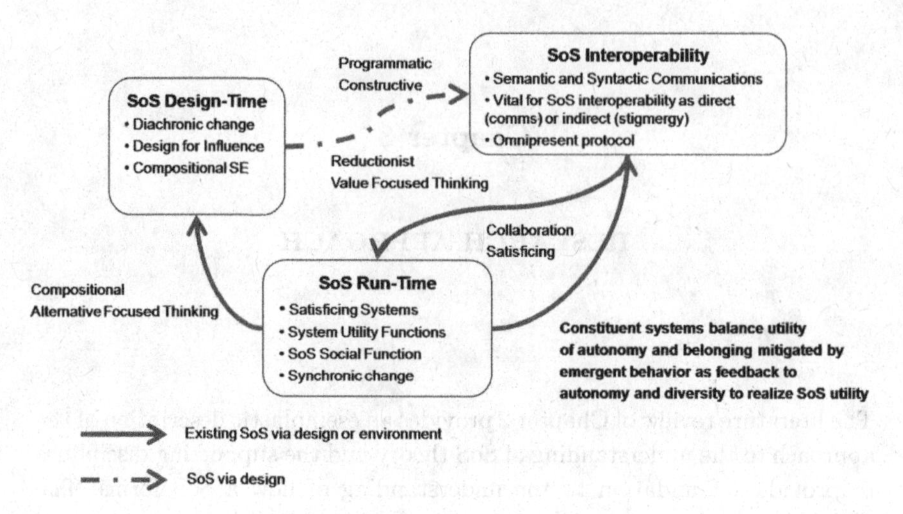

Fig. 1. Research Approach.

& Meyer, 2001; Garnier, Gautrais, & Theraulaz, 2007; Winans & Bassler, 2002). In order for any communications to take place whether it is direct or indirect, there must be a protocol present for the respective systems to be able not only connect in a form of communications but also be able to understand the communications. The aspects of syntactic and semantic communications respectively to support this idea are included in the approach.

3.2 Systems of Systems Design-Time

For both design-time and run-time SoS, the research approach is of a constitutive nature versus the summative. Summative nature is assumed to be traditional systems engineering via reductionist engineering — an engineering approach that provides decomposition from system requirements to design, construction, and validation. The constitutive approach is compositional, herein referred to as compositional engineering, in nature whereby systems are engineered for influence or a design for influence from existing legacy systems or the addition of new systems within a legacy environment.

The nature of SoS assumes an important distinction from traditional systems that a SoS is a system but not all systems are a SoS. This is based on the idea of holons found in (Koestler, 1967) and further refined by (Boardman, 1995) is shown as a spectrum against the SoS characteristics

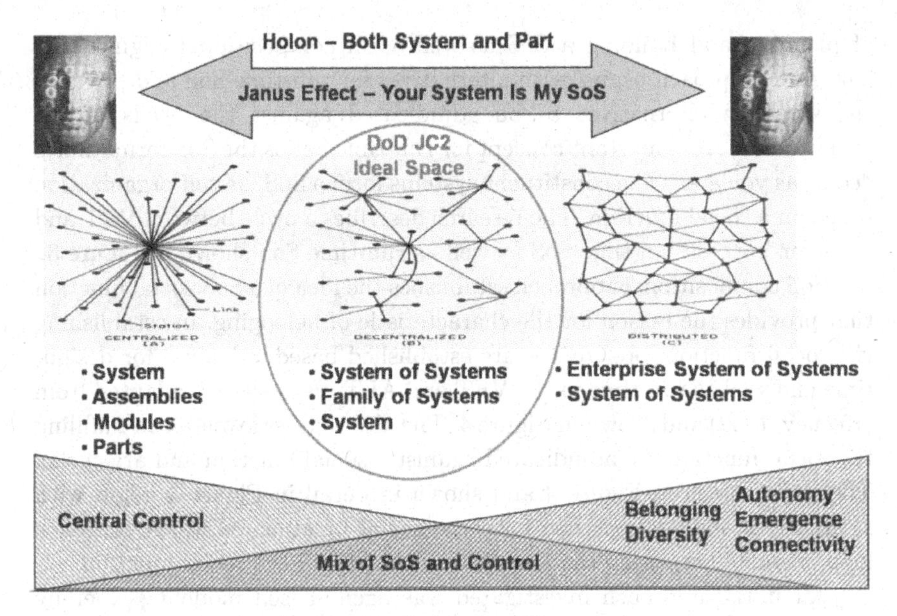

Fig. 2. SoS Janus Effect.

in Figure 2 in regard to the DoD Joint Command and Control (JC2). This approach reveals that the SoS characteristics are able to have variations whereby they may be “dialed” to exhibit variations of the behavior of the system collective that provide for the SoS. The variations are due to spatial placement in the hierarchy of systems whereby the observer is not at the same level as the systems being observed to understand how they are organized to fulfill the whole (Driscoll, 2008). The idea shown in Figure 2 is the DoD is most likely comfortable with a hybrid of systems that are distributed than with an environment with no command and control and those systems that are entirely within a strict command and control structure. This is an extension of Paul Baran’s work of distributed networks whereby a distributed network provides greater capabilities and survivability (Baran, 1964).

3.3 System of Systems Run-Time

Run-time SoS uses the concept of Value Focused Thinking (VFT) and Alternative Focused Thinking (AFT). The VFT establishes within a hierarchy the values and goals of subtending nodes in a reductionist approach. This approach is important for design-time SoS as a function of planning and building a SoS as well as in compositional engineering. The AFT approach begins with alternatives for run-time SoS and provides the available alternatives to be adjudicated against the needs of the constituents. An important concept for run-time SoS is the SoS formation is "come as you are." The constituent systems form a SoS via self-organization based on a social function. The research describes a cycle between VFT and AFT for both design-time SoS as well as run-time SoS shown in Figure 3.

SoS compositional approach establishes the idea of a SoS social function that provides the reason for the characteristic of belonging. In establishing the social function, preferences are established based on values for design-time and well as run-time in the VFT and AFT methodology adapted from (Keeney, 1992) and shown in Figure 4. The alternatives available in fulfilling the social function are adjudicated against a social function and are cyclic. The cycle shown in Figure 3 and shown in detail in Figure 4 begin with alternatives and are matched with the social function values. A feedback loop is created selecting the alternatives that fulfill the social function.

An initial approach investigated was agent based modeling whereby agents represented constituent systems. Ontological engineering was investigated to articulate how the agents would communicate to form a SoS (van Diggelen, 2007; Williams, 1999). This research investigation approach was abandoned due to the lack of visibility of how agents in available agent software actually communicate. As a result of the agent based ontological engineering approach, a SoS mechanism based decision theory approach is investigated.

The SoS mechanism describes the social function that enables systems to cooperate fulfilling their utility function. The environment may be

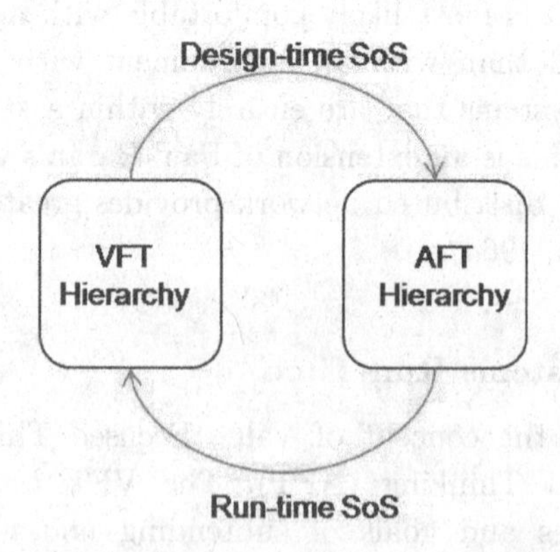

Fig. 3. SoS Compositional Hierarchy Circular Process.

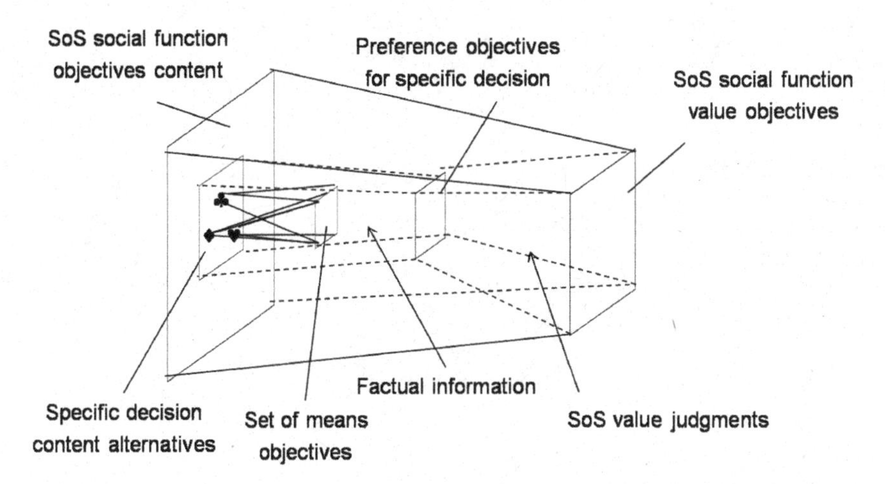

Fig. 4. SoS Social Function Value Thinking (Keeney, 1992).

unidirectional or bidirectional whereby the social function and mechanism design rules and policies are modified per changes in the behavior of the constituent systems. The mechanism consists of two aspects adapted from (Russell & Norvig, 2003):

1. Ontology for describing the set of allowable policies that the constituent systems can adopt, and
2. The resultant rule that determines payoffs to the constituent systems given the policy profile of the allowable policies.

In cooperative systems, the constituent systems are allowed to make binding agreements. It is assumed that the agreements have been made over a range of possible alternatives, which is the basis of bargaining theory and is not detrimental to them. A SoS mechanism implementation creates the social function executing the allowable policies that imparts to utility theory the reasons for systems to satisfy their EU. The mechanism may be as simple as a voting scheme whereby systems vote if certain rules are satisfied and the winner is determined against another set of rules.

Chapter 4

SYSTEM OF SYSTEMS FORMATION

4.1 System of Systems Design-time and Run-time Overview

SoS interoperability (SoSI) plays a fundamental role in the creation and management of a SoS (DiMario, 2006) because it can only exist due to the communications of constituent systems, syntactically and semantically. SoSI is discussed at length in Appendix D. Interoperability is also at various levels with various ontologies depending upon the framework that interoperability exists. The independent systems become interdependent through integration via interoperability forming a relationship of mutual dependence and benefit between the interoperable systems. Both systems and SoS conform to the accepted definition of a system because each consist of parts, relationships, and a whole that is greater than the sum of the parts. However, the spectrum of systems via the Janus principle reveals that a SoS is a system, not all systems are a SoS.

The SoS Janus Principle hierarchy applies to SoS as a system as it does to a non-SoS system and its components or parts producing a hierarchy of a SoS. It is reasonable to assume that at very high hierarchical levels there are multitudes of SoSs that interoperate with one another. For example, the National Transportation System is a SoS comprised of several SoS as airlines, trains, and highways. Each of the SoSs that comprise the SoS National Transportation System is lower in the hierarchy with other systems comprising their SoS and so on. A differentiator in the Janus hierarchy is the division between independent systems and SoS. For the independent systems, their capability is the summation of their parts. For the SoS, their capabilities are the result of interdependent systems providing a nonlinear dynamic interoperability resulting in capabilities greater than the sum of their parts.

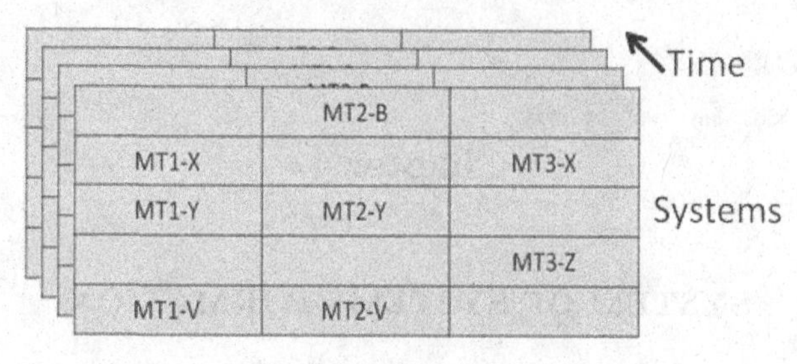

Fig. 5. SoS Network Centric Operations.

There are many perspectives of the systems that is a SoS that are dependent on the system user, stakeholder, or customer of a system's product or service. This is shown specifically in an example of DoD Network Centric Operations (NCO) shown in Figure 5 whereby there are a myriad of disparate systems for various types of mission threads or NCO capabilities. Mission-Thread 1 (MT1) will use three distinct systems MT1-X, MT1-Y and MT1-V and Mission Thread 2 (MT2) will use MT2-B, MT2-Y, and MT2-V respectively. These systems are designed to work together to perform a specific type of capability and called design-time SoS (Carney, Fisher, Morris *et al.*, 2005). Once more, these systems have their own life cycle that changes with time. System MT1-Y may change without any knowledge of nor care about the systems in Mission Thread 1 and will move to a different temporal plane via temporal system's change, called diachrony and discussed in the section of emergence, leaving to chance the interoperability of the other two systems. The systems in a mission thread may comprise a radio, radar, or a command center. The SoS that may arise is also based on immediate needs. For example, a military service unit finds itself in a new environment with a new mission requiring an unexpected capability; the new mission thread may consist of systems MT1-V, MT3-Z, and MT2-B cobbled together by a run-time SoS process. These systems may not have full interoperability as if they were designed for efficiency, but the user has already determined an acceptable level of interoperability that is satisficing based on needs (Gorod, DiMario, Sauser & Boardman, 2009) discussed in Appendix F.

There is absent a common lexicon to describe or understand the relationships and differences between design-tine and run-time SoS. The

Table 4. SoS Formation Boardman Conceptagon.

Conceptagon	SoS as a Result of Design-Time	SoS as a Result of Run-Time
Function structure process	Composing non-autonomous constituents or predefined autonomous systems; Hierarchical	Composing autonomous constituents; Flat; Distributed
Relationships wholes parts	Predefined at T_0; Typically reductionist and VFT methodology; Utility functions predetermined	Loosely predefined to undefined; Composeable; AFT methodology; Systems determine utility function levels
Emergence hierarchy openness	Predicted for T_0	Unpredicted; Evolutionary
Communication command control	Central to decentralize; predicted and designed omnipresent protocols	Central to decentralize; predicted and designed omnipresent protocols; Amorphous command and control
Boundary interior exterior	Fixed to Dynamic	Dynamic to Adaptive
Transformations Inputs Outputs	Predefined and flexibly bounded	Loose and undefined
Harmony variety parsimony	System to predetermined social utility function	System to undetermined social utility function

Boardman Conceptagon in Table 4 summarizes a design-time versus run-time SoS as a set of seven triples representing the systems engineering lexicon that provides a uniform language of disparate systems of interest (Boardman & Sauser, 2008; Boardman, Sauser, John, & Edson, 2008). The comparison reveals a spectrum of fixed to dynamic behaviors for design-time SoS and dynamic to distributed behaviors for run-time SoS.

4.2 SoS Characteristics and Evolution

The following SoS characteristics are discussed in detail as they pertain to SoSI, design-time SoS, and run-time SoS as general approaches. The subject of evolution is added to the discussion of the SoS characteristics because of the amount of influence evolution has on the nature of a SoS. The concept of SoS evolution is arguably captured in the characteristic of emergence and is discussed later in terms of feedback to the constituent collective. However, it has a strong significance and is treated as a sub characteristic to emergence. A finer discussion of the characteristics and how they may distinguish between design-time and run-time SoS are treated in their respective topic discussions and case studies.

4.2.1 *Emergence*

A foundation of understanding the characteristics of emergence is that of change. If systems are static in either themselves or their environment, there would be no change and because of no change, there would be no emergence. The basis of systems theory is:

- The foundation of a system is determined uniquely by its structure, and
- The response of a system is determined by its function and the history of its inputs.

Change is represented in this thesis by the concepts of synchrony and diachrony, borrowed from social systems analysis (Cortes, Przeworski, & Sprague, 1974). Synchrony is the stasis of a system of its structure and function. The structure and function do not change and are not subject to the effects of time and change in itself, its inputs or its environment. Synchrony must be continuously reproduced for if anyone of these changes, the system ceases to exist. Synchrony tolerates dynamics as temporal variation of its inputs, outputs, and states as long as variation is within the system's synchronic boundaries. The dynamics of a system's response is determined by its synchrony. Synchrony represents the stasis of a system while diachrony represents changes of a system's structure and hence the system itself. A new structure appears and the nature of the interrelationship process changes. It is this diachronic change whereby systems evolve because of survivable emergence and transition through synchrony.

Patterns or capabilities in a SoS through interoperability will manifest through system operation. SoS emergence originates based on the following:

- For man-made and natural systems, SoS capabilities or SoS behavior is the result of an interaction that is unplanned and may result in beneficial or unintended consequences. This is entirely based upon the needs of the observable. For natural systems, emergence may result from cooperation, competition, and evolution. Within an ecosystem, asymmetric species compete for resources such as food. Species may also cooperate in symbiotic relationships in a symmetric interaction. Ecosystems are determined by the environment but also from interaction of the organisms. Evolution, through selection and variation, drives diversity and success of species through competition comes from interactions of asymmetric species. Likewise is the case for man-made systems. As asymmetric systems evolve due to changes in their system life cycle or in the case of genetic algorithms (Holland, 1995; Mitchell, 2001), their interactions will determine their survival.

- System level attributes are irreducible in that they cannot be explained or inferred from the structure of the cooperating constituent systems or their properties.
- System level attributes are not held by the constituent systems.
- System level properties can only exist at the system level.
- System level effects can only be understood through composeability from the lower levels of the system.

Emergence is viewed as being a property of the whole system and not subject to its parts and is unpredictable. It is also said that emergence is perfectly predictable. Both perspectives are true because it is based on the relationship of the system and its observer. It is emergent or unpredictable to one observer and predictable to another. If emergence is taken as an absolute view, then it would be a surprise (Weinberg, 2001). Based on the two perspectives of emergence noted, emergence can be engineered with predictability or appear unpredictably with unintended consequences resulting in benefits or cost to the observer.

In complex adaptive systems, emergent behavior is the result of elements producing a behavior that resides on a level above them. Emergence is the movement from low-level rules to higher-level rules. The elements follow local rules and through their local interaction results in a macro behavior. A higher level pattern is the result of parallel interactions between local agents (Johnson, 2001). Lower level elements possess mutual relationships or a reason to belong whereby they have no reason onto themselves. Through this relationship, an emergent behavior results at the next level and is defined by the relations of the observables or aggregate property of the elements at the lower level (Herbert A. Simon, 1996). Emergent behavior is not considered emergent until it is manifested in a macro behavior while emergent complexity without adaptation is similar to snowflake crystals whereby it is a complex orderly pattern without function.

It is difficult to predict the outcome of emergent behavior and how the effected systems will adapt. What is certain is that systems will seek and maintain preferred states and interactions are rewarding, not only based on the interaction but the interdependence and resulting macro behavior. The mechanisms of adaptation in collective interdependent systems are exhibited by four principles (H. Arrow *et al.*, 2000).

1. The principle of non-proportionality — effects are not governed by the size of the effect or event. Large events may result in no response whereby small events will result in huge changes. The effects are essentially nonlinear.
2. The principle of unintended consequences — when complex systems interact, the resulting behavior is difficult to predict. When changes are introduced, the resulting behavior may be beneficial or detrimental to the observer.
3. The principle of temporal displacement — systems are changed to accommodate the needs of the future but make changes with no basis of the current configurations. Systems respond to events that do not take place in response to poor planning.
4. The principle of spontaneous innovation — small groups of systems, a natural collection of systems, or adjacent systems will act as a closed system introducing new routines and local behavior. However, this is an open system resulting in unintended consequences and will persist if the rewards are greater than the rules replaced.

For autonomic computing systems and a goal of computer science is to build massively parallel-decentralized computing systems, the mapping from local behavior to global behavior is nontrivial due to its fundamental properties of self-configuration, self-maintenance, self-optimization, and robustness (Ganek & Corbi, 2003). Designing behavior at design-time at the local level is insufficient to derive behavior at the global level due to non-linearity of interactions. Techniques have been developed to map local evolution rules using genetic algorithms and cellular automata to achieve global behavior. This approach is modeled as to how the neural systems of the brain work in a collective manner and to design computer systems in such a manner including evolution using genetic algorithms (Crutchfiled, Mitchell, & Das, 2003; Mitchell, 2001).

Interactions of the constituent systems are non-linear so the behavior of the system cannot be obtained merely by summing the parts or by direct inspection of the rules of the individual systems. Macro rules can be formulated by observed patterns based on the whole system without resorting to micro rules of the individual systems (Holland, 1998). The emergent behavior associated with elements at a particular level constrains the degree of freedom of the elements. An example is the arrangement of deoxyribonucleic acid (DNA) bases, the fundamental basis of life, follows the laws of physical chemistry. The arrangement places constraints upon chemistry which produces the property of genetic coding. The genetic coding may be an emergent property that marks the transition from the level of chemistry to that of biology. Any arrangement of the organic bases is compatible with the laws of physics and chemistry (Checkland, 1993). The proper arrangement of the organic bases for genetic code is critical.

A SoS health care infrastructure example to consider consists of systems that comprise complex autonomous and heterogeneous systems. These systems form partnerships whereby their interoperability relationship is complex and adaptive, producing emergent behaviors that do not originate from anyone individual constituent healthcare system. The healthcare SoS is subject to resistance to change due to dynamic complexity whereby new behaviors, or unintended consequences, arise from the interoperation of the constituent systems over time. Dynamic complexity is not subject to conventional forecasting, planning, and analysis methods (Midgley, 2006; Senge, Kleiner, Roberts, Ross, & Smith, 1994; Sterman, 2006). An example of an unintended consequence is reducing the amount of tar and nicotine in cigarettes, which then causes smokers to smoke more as they compensate for the lower tar and nicotine. Another example is the increased resistance of certain strains of bacteria such as Staphylococcus aureus and sexually transmitted diseases due to the overuse of antibiotics. The unintended consequences are due to not understanding the complexity of the interoperability relationships of the constituent systems. The relationships are either independent or interdependent partners in a holistic result. The emergence is unpredictable because the emergent pattern is not understood and thus the observer has an unintended consequence. Once understood the emergent behavior is entirely predictable and may be resolved through properly engineering the infrastructure.

True autonomous constituent systems are capable of independent action with limited influence from outside forces. Influence is any mechanism by which an agent interacts with another agent in such a way that the interaction changes the state of the other agent. It is a fundamental mechanism for all interactions among autonomous entities. By increasing their influence, agents can achieve goals that are not local to themselves through cooperative interactions. The mechanisms of influence consist of cascade effect *s* and emergent composition.

In practice, there is no doubt that there is great synergy from disparate systems if they are able to interoperate for the good of the whole. Man-made technological disparate autonomous systems that interoperate do not naturally display characteristics that are greater than the sum of their constituent systems as they were designed to cooperate and their holistic behavior is entirely reducible. The arising holistic capability and emergence may be reversely decomposed to their constituent systems and subsystems. For a phenomenon to be called emergent, it should be unpredictable from a lower level description and may be irreducible. The frequency of complex interactions may not derive emergence. It is not the number of interactions, which encourages emergent behavior, but how they are organized such as neurons in the human brain. In some cases, the system must reach a threshold or tipping point of diversity, organization, and connectivity before emergent behavior appears.

The influence of a SoS and subsequent emergence is by means of cascading effects resulting in local interactions which in turn create new properties that are not derived from simple summations of the SoS constituent system properties (Fisher, 2006). A special form of a cascade effect is an epidemic that does not have a natural tendency of dampening. In this effect, the numbers of constituent systems that are influenced increase at each transition. This is similar to epidemics of disease. However, all epidemics are eventually halted due to organized resistance, resource limits, or a saturation of possible systems or the removing of systems in the chain. Systems thinking archetypes have similar cycles called amplifying or stabilizing loops. A reinforcing loop incrementally changes the state of the constituent systems. Reinforcing loops create a spiral of effects that result in success or failure. However, they cannot continue for long as they result in creating balancing loops and are consumed by them resulting in Limits to Growth archetype. A Limits to Growth archetype describes situations whereby growth cannot continue without encountering a limiting factor (Senge *et al.*, 1994). For SoS, the composition effect is a fundamental attribute by which autonomous systems are combined to produce a configuration that cannot be expressed as a summation of the constituent systems. The emergent property that arises tends to be global as there is no single system or groups of systems that produce the emergent property, but their collective interoperation.

A collective emergent interoperation may be designed such as the NASA Autonomous Nanotechnology Swarm (ANTS) program being considered for the 2020 period. The proposed ANTS mission is comprised of 1000 disparate autonomous pico-class (approximately 1 kg) satellites that will explore the asteroid belt. Approximately 80% of the satellites will have a single specialized instrument onboard and obtains a specific data type. A few satellites will be coordinators and others will be messengers that will coordinate communications between the specialized satellites, coordinator satellites, and Earth. The swarm must be verified for their proper collective behavior and that no undesirable behaviors emerge. There are a large number of concurrent interactions taking place among the pico-satelites. Verifying the swarm is difficult due to complexity of each member and the complex interaction of a large number of intelligent elements (Truszkowski, Hinchey, Rash, & Rouff, 2006)

An example of emergent behavior in nature is a possible deep freeze of North America and Europe due to global warming. The effects of extremes, warmth to cold in order for nature to correct the effects of rising carbon dioxide levels fuels the debates of global warming. Portrayed in the 2004 movie "The Day After Tomorrow" and discussed in the 2006 documentary film "An Inconvenient Truth" North America and Europe enter Antarctic like conditions due to the halting of the Atlantic Ocean current known as the conveyor belt. The conveyer belt transports warm surface water from lower latitudes to the northern waters and carries colder deeper water to the equator. The conveyer belt process moderates northern temperatures. The conveyer belt is disrupted due to an increase of fresh water from melting snow, ice, and glaciers of the arctic and Greenland. As the salt-water density decreases, the transport mechanism also decreases thus halting the conveyer belt. Once the belt halts, the Northern hemisphere is plunged into Antarctic conditions, as warmth from the south is no longer present.

For our air traffic control infrastructure, an example of emergent behavior that results from the operations of the air traffic control system is the slowing of departures as a day's operations are carried on. Due to inefficiencies in a highly distributed air traffic network with occasional storms that cause delays that eventually ripple through the network is not the result of any one system or the summation of the systems. It is resultant behavior of the entire collective system. There is a tipping point whereby an accumulation of effects can no longer be absorbed by buffering mechanism in the transportation SoS and will have a profound effect upon the entire system.

4.2.2 *Autonomy*

The literature characterizes autonomy as primary and essential to define a SoS as it is a differentiator of system participants of a unitary system or a SoS (Boardman & Sauser, 2006; DeLaurentis, Fry, Oleg, & Ayyalasomayajula, 2006; Maier, 1996). There is also a degree of autonomy or independence that must be considered or how independent is independent? Does a semi-autonomous constituent system within a SoS constitute a characteristic for SoS? For example, the varied systems of the International Space Station are autonomous depending on their usage at a point in time and how tightly coupled the constituent systems are to other independent systems. This leads to misclassifications of systems as either being unitary or a SoS. This is rooted in the myriad of perspectives of autonomy. Autonomy has a spectrum of perspectives from an absence of autonomy to semi-autonomous to fully autonomous. But what is autonomy?

Autonomy means independence and self-governing. A framework of autonomy levels is suggested for unmanned systems based on the amount of human independence (Huang, Pavek, Albus, & Messina, 2005). For purposes of this thesis, autonomy includes autonomicity, which is the ability to exhibit autonomic properties. Autonomic properties of systems are self-configuring, self-optimizing, self-healing, and self-protecting (Ganek & Corbi, 2003; Truszkowski *et al.*, 2006). For instance, in autonomic computing, computer systems manage themselves given high-level objectives through established policies. These systems are self-governing autonomous systems that interact with other autonomous systems. When a new component or system is introduced, into an autonomic system, it automatically learns about and takes into account the composition and configuration of the system. It registers itself and its capabilities so that other components can either use it or modify their own behavior appropriately. For the ANTS program, autonomy alone is insufficient even if there is no human intervention as the satellites will be vulnerable to the harshness of space and with very little or no communication with their creators. ANTS spacecraft engineered with autonomic properties will ensure their continued survival and mission success.

4.2.3 *Diversity*

Diverse or heterogeneous constituent systems may characterize a SoS due to the connectivity and interoperation of diverse capabilities leading to emergent behaviors. If there were no diversity, there would be no reason to interoperate except for improved resources or reasons to have more of the same. Diversity drives the potential for emergence through configurations of connectivity and evolving constituents. Diversity provides the capabilities that are greater than the simple summation of constituent systems through collaboration and a sense of belonging to fulfill a need.

4.2.4 *Connectivity and Interoperability*

Connectivity is the interoperability of the constituent systems expressing their requirements and needs. Connectivity may comprise any means necessary to link and to send information between constituent systems of a SoS. The internal machinery of the receiving system makes use of the information for the benefit or detriment of the SoS. In order to be persistent and common, connectivity may range in composition from chemical communication of bacteria (Bassler & Losick, 2006) to networks for computers and society. The connectivity characteristic is an interoperability enabler (DiMario, 2006). A definition used in this context is: "The ability of a collection of communicating entities to (a) share specified information and (b) operate on that information according to a shared operational semantics in order to achieve a specified purpose in a given context" (Carney, Fisher, Morris *et al.*, 2005). Connectivity and interoperability is discussed at length in Appendix D.

4.2.5 *Belonging and Collaboration*

Interoperability is dependent on the characteristics of the constituent systems as well as the nature and interaction of other characteristics of other systems. In SoS, thus follows that the state of the SoS as a system is determined by its history of interactions that have occurred thus far. Systems follow their own interest and the resultant system formed via interoperability is based on the individual objectives of the constituent systems and the behavior of the whole can be only estimated at best via deduction. What is important for SoS is to discover what is necessary for cooperation to emerge, appropriate actions can be taken to foster development of cooperation in a specific setting. Cooperation is based on an investigation of the individuals who pursue their own interest without aid of a central authority to force them to cooperate. There are certain cases in which cooperation is not completely based upon a concern for others or upon the welfare of the group as a whole as demonstrated in the popular iterated prisoner's dilemma (IPD) (Axelrod, 2006). The IPD is a two-player game whereby players try to maximize their performance or fitness by cooperating with the other player or defecting. The payoff received is dependent on the actions of the other player. Axelrod set up experiments in which strategies were evolved using as their fitness the mean score attained against a set of eight human strategies. He illustrated that the system evolved the best strategy.

The analogy to demonstrate cooperation and a sense of belonging is taken from biological organisms as mutualism, which is an association, as symbiosis or parasitism, of two organisms. The cooperation of organisms for a mutual benefit is symbiosis and a negative effect is parasitism. Parasitism results in the death of both organisms whereby symbiosis results in coadaptation and coevolution. Symbiosis can also result in endosymbiosis whereby two species become so interdependent that they become inextricably linked. This is akin to different parts of a problem are so interdependent that they are not amendable to division. The symbiotic analogy extends to systems as organisms. A system is said to be endosymbiotic when it cannot be separated or decomposed and yet fulfill its intended purpose and loses its autonomy. A system is symbiotic whereby it cooperates and shares its capabilities and resources with another system to fulfill a mutually agreed upon need. A system that is interoperable is symbiotic and maintains its autonomy.

4.2.6 *Evolution*

All useful systems evolve, but in traditional unitary systems, evolution has seldom been treated as an integral aspect of the design, implementation, management, and operational process. In a SoS, the management and operational independence of the constituents enables their independent evolution. This independence of change in individual constituents adds significantly to the complexity of the interactions among constituents and of management and operations. Thus, in SoS, evolution must be explicitly recognized and managed. Explicit recognition encourages use and exploitation of evolution and, therefore, more frequent changes. Even without increased frequency of change in individual constituents, evolution will appear more continuous from a global perspective due to the lack of system-wide coordination of evolutionary changes.

Because almost all systems evolve in response to changing needs and technological advances, the fact of evolutionary development alone cannot distinguish a SoS. A SoS is unique in that its autonomous components can evolve independently of one another. Without knowledge of how their neighbors are evolving, constituents are likely to evidence incompatibilities, with unanticipated favorable or unfavorable effects relative to the observer. This creates a level of complexity in the evolution of systems not found in unitary systems. Given the context of evolution for SoS versus unitary systems, the motivation for evolution is the similar for both. However, the impact for the SoS is largely different from that of unitary systems.

Several definitions of SoS evolution have been posited such as (Carney, Fisher, & Place, 2005; P. Chen & Clothier, 2003):

1. Preservation evolution whereby the SoS preserves functionality during the evolution of its constituent systems, and
2. Adaptive evolution whereby the SoS has evolved due to constituents that have changed or there are new constituents.

On the surface, the definitions appear to be simplistic until one begins to unravel the complexity of SoS evolution. Emergence and evolution are tightly coupled with emergence as a predecessor to evolution. Emergence must appear as a macro behavior for the collective interacting systems. Evolution is based on change that leads to greater complexity as a function of the environment as decentralized systems generate structure due to synchronic and diachronic change.

Real networks are controlled by two laws of growth and preferential attachment. A network starts with a small node and expands with additions of new nodes. The new node, when deciding to link, connect with other nodes that have more links (Barabasi, 2003). Case in point are the formation of cities whereby a small group will form and settle in a particular area for a particular reason being available water, goodness of the soil, weather, waterway transportation, and a host of other reasons. The original group had no specialist skills such as hunters, agriculturalists, blacksmiths, or other varied skills. All members have the same relative skills until sufficient growth begins and the community experiencing a greater need for specializations. Collections of specialists are generated as well as groups within the city such as neighborhoods over time. Patterns in time become dominant and are self-organizing. Neighborhoods absorb change through a process of homeostasis whereby they modify their processes and reach an internal equilibrium before the city at large feels the effect. A complex system-defining characteristic is that global behavior outlasts any of its components. Generations of bees live and die supporting the colony through their specializations. The colony matures and grows in self-organizing patterns. The same is true for a city.

The evolution of a SoS open system eventually leads to self-differentiating systems that grow to greater complexity. Complex adaptive systems are special cases of complex systems. They are made of large numbers of active elements that are diverse in form and capability. They are adaptive in that they have the ability to change and learn from experience. The behavior depends on the interactions of the constituent agents of the system or network. The aggregate systems may again be aggregated to add new hierarchical systems (Holland, 1995). Adaptation is described by Holland as the ability to learn and relate processes with many varying over different degrees of time via sets of rules that form patterns. The systems adapt by changing their rules as they gain knowledge through time. However, not all complex systems are SoS and not all SoS are viewed as complex. But, it would seem that there are similarities between complex systems and SoS. Complex adaptive systems examples are cited as the stock market, ant colonies, the biosphere, the brain, and manufacturing businesses.

There are two suggested forms of group adaptation — undirected and directed (H. Arrow *et al.*, 2000); which seems suitable for SoS. Undirected adaptation is the result of small changes in the environment, not part of any planned scheme, but as normal variation. This normal variation can be change associated with iterations of other systems in symbiosis relationships. Behavior of systems will be modified to accommodate the changes and will persist of the rewards are greater than the costs. Over time, the aggregation of changes based on positive feedback and negative reinforcement will alter the group's global structure and macro behavior. This is known as undirected emergent behavior over time resulting in classic evolution. This is discussed later in this thesis in the terms of alternative focused thinking (AFT) and run-time SoS.

The other form, directed adaptation, is goal oriented whereby new behaviors and structures are determined by a plan. The plan can originate internally or externally to the systems. An example is a global change to NCO as to how transmissions between large communication hubs are performed. All systems or nodes will uniformly adjust or lose their ability to communicate or be interdependent. In a SoS another means is design to influence which is "point engineering" to "influence engineering" whereby the objective function for my own system design may influence the other systems in such a way that interoperation fulfills my objectives.

While systems are autonomous, they can be influenced and not controlled. One system interacts with another in a way that changes the physical, informational, or end state of the other; whether it is positive or negative is the judgment of the observer. This is discussed further in this thesis as Value Focused Thinking (VFT) and design-time SoS.

In both forms, adaptive coevolution of interdependent systems exists whereby they adapt to one another in a series of reciprocal changes. This is done in mutual reinforcing feedback loops. In a predator-prey ecosystem, a pseudo arms race is evoked whereby improvements in prey defenses evoke improvements in predator offense, and vice versa. This results in a global macro behavior change of the ecosystem or SoS.

4.3 Design-Time SoS Formation

4.3.1 *Enterprise Architecture*

The concept of design-time SoS is focused on the SoSE of a SoS whereby a group of systems may be planned and architected to have an influence upon one another, thus design for influence. This avoids the large integration of autonomous systems into one large system that produces inflexible interfaces and expensive development and testing cycles. However, the converse produces many stakeholders and owners of the myriad of systems that comprise the SoS. Appendix E discusses the use of the Zachman Framework (ZF) modified to architect a SoS.

A major challenge is to organize and manage the information of the myriad of stakeholders and requirements. This approach is at the enterprise level whereby it is compositional methodology versus a reductionist methodology typically performed in traditional systems engineering. The ZF provides the various artifacts to describe the architecture from several perspectives. It modified and clarifies the notion of "... your system is my SoS..." providing a contextual basis among stakeholders with different levels of enterprise perspective (DiMario *et al.*, 2008).

4.3.2 *Composeable SoSE*

Design-time SoS or also referred to as composeable systems engineering in this thesis has several components:

1. Design-time SoS at T_0
2. Design-time SoS at T_{0+1}
3. Run-time SoS producing the design-time at T_{0+1}

Design-time SoS at T_0 is a "green" start SoS whereby the entire SoS architecture is planned and constructed from high level requirements through design and validation. An example is the U.S. Coast Guard Deepwater Program comprised of a myriad of ship assets, aircraft, and supporting systems (Acquisition Solutions, 2001). This SoS government program example is indeed rare and thus this research does not address it since it is a traditional systems engineering process. However, at some time epoch noted as T_{0+1} at least one of the constituent systems of the Deepwater Program, for example, will begin its own life cycle independent of the other constituents. All the constituent systems were originally planned, organized, and constructed in uniformity at T_0. But due to distributed governance of the individual systems, the individual systems begin to have a "life of their own" (U.S. Department of Defense, 2008).

Design-time SoS at T_{0+1} manifests itself in the aforementioned example but also appears in a run-time context. A simple example is when a warfighter solves a problem by creating an interoperable method between two systems that were not designed to work together in the field by moving data from one system to the other with a memory stick. The warfighter's problem required a new capability and it was solved by the realization of moving data from one system to another. In the context for this particular warfighter, the problem may never be seen again as it exists per a specific environment for a specific moment in time. The creation of a new capability from two existing systems is a simple example of a run-time SoS leading to a composeable engineering design-time of T_{0+1} for the respective systems and shown in Figure 6. The products of SoSE enter a cycle of design-time reductionist engineering and at some time later, the constituent systems recompose themselves via self-organization

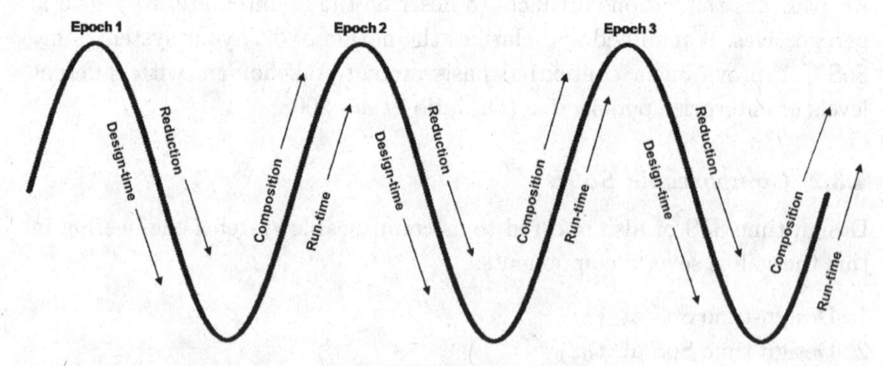

Fig. 6. SoS Epoch Composeable Engineering.

or real-time "redesign" due to changes in the environment (synchronic change) or needs of the users who identify value in the evolution of the constituent systems creating a new level of constituent system belonging in context of the SoS. The compositional engineering is the "reassembling" of independent systems to work together in a holistic manner. The user who finds value in the constituent system's mutual belonging defines the system's interdependence.

4.3.3 *Composeable Design-Time SoS Case Study*

A case study and subsequent SoSE process was created base on Close Air Support (CAS) example cited in (Sund, 2007) defined as "Air action by fixed and rotary-wing aircraft against hostile targets that are in close proximity to friendly forces and that require detailed integration of each air mission with the fire and movement of those forces" (U.S. Department of Defense, 2001). The problem presented is a result of introducing interoperability into a mix of new and legacy systems. To affect a CAS in different warfighting environments and mission challenges, there is a mix of systems per mission threads represented as a SoS Environmental Cube (E3) in Figure 7. The SoS E3, a functiòn of time, systems, and capability, is environment and temporal specific as the operational capability MT1 in Afghanistan may be much different from a MT1 in Iraq. The difficulties represented by the SoS E3 are that each system has its own life cycle and

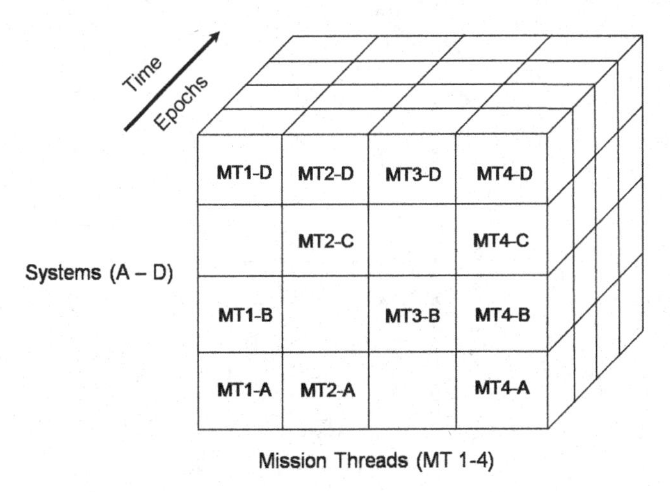

Fig. 7. Systems of Systems Environmental Cube (E^3).

managed independently of all other systems. Each system as well as their governance has very little if any view of the other systems that are used in a mission thread. As a function of their individual life cycles, systems evolve as a function of time at different rates.

The SoS Cube decomposes the mission threads into a SoS sub-E^3 representing the Command and Control (C2) net-centric systems of bandwidth capabilities, network capabilities, services, and applications versus warfighting functions shown in Figure 8; the warfighting functions are:

- Fires — artillery support;
- Maneuver — kill chain timeline; movement of friendly forces to take advantage of enemy forces;
- Intel — Intelligence; collection and dissemination of information;
- Log — Logistics; management of resources;
- FP — Force Protection; protection and purposeful lowering of risk to friendly forces.

The operational capability exhibited in each mission thread as a function of the collection of systems is a matter of perspective of those that find value in the collective systems of interest that form a SoS. Using the ZF aspects of Who, What, and How discussed in Appendix E, are matched to the organization, the required data, and the function respectively while

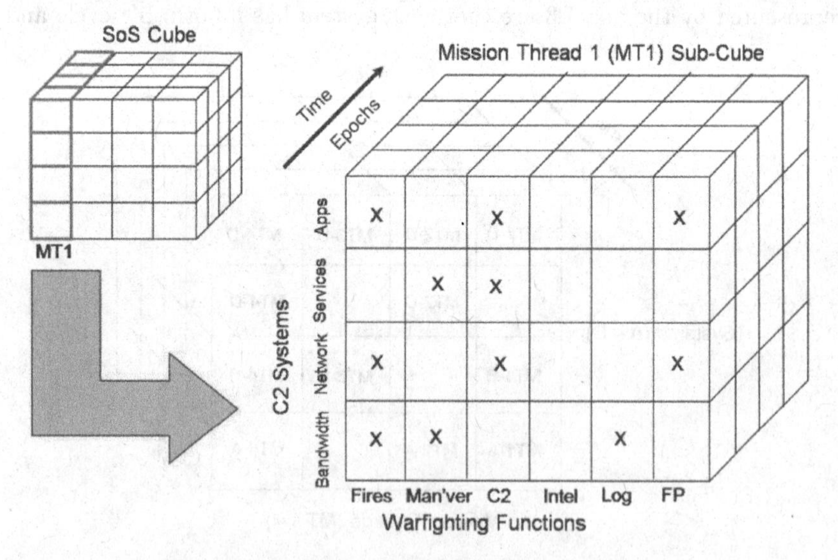

Fig. 8. SoS Environmental Sub-E^3 Decomposition.

the other aspects of What, Where, When, and Why have a tendency to remain static for a specific mission thread at a specific epoch. This matching is required to realize the collective systems of an operational capability mission thread.

4.3.4 *SoS Design-Time Compositional Process*

The process shown in Figure 9 is a combination of a reductionist approach of steps one and two and compositional for steps three through five. This process identifies functions to systems gaps as well as excess systems and systems capability gaps by aligning functions to capabilities. The objective in compositional SoSE is to enable capabilities through the constitution of systems and their functions by maximizing functions to systems mapping and minimize excess systems. The steps one through five are shown in detail in Appendix A. Step one is concerned with populating a database of systems functionality per mission thread per epoch. Step two creates use case diagrams and mission threads. Step three evaluates the mission threads and creates mini-threads that represent patterns as a compositional process and fulfilling the ZF approach of identifying Who, What, and How. It is the patterns that become the composeable engineering for a mission thread as well as across mission threads and epochs are the common denominator. Step four consolidates the mission threads for common syntactic interoperability for detect, control, and engage functions.

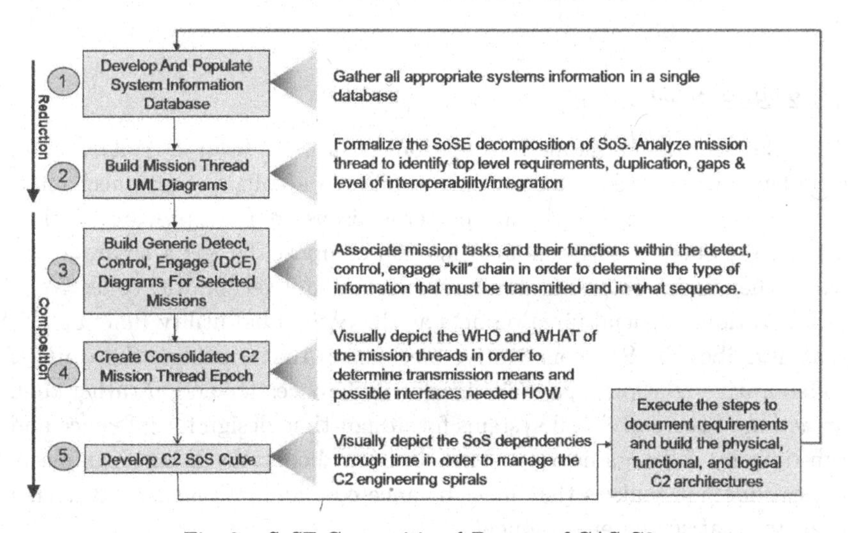

Fig. 9. SoSE Compositional Process of CAS C2.

Step five aligns the mission thread epochs to programmatic views adapted from the Ministry of Defense Architecture Framework (MODAF) (MODAF Development Team, 2008).

Analysis of the entire mission reveals repeated patterns or mini-threads. A mission can be composed of multiple mini-threads. Often used tasks performed by multiple platforms highlight where commonality of interfaces facilitates integration or interoperability. Analysis of data transfer and dependencies over multiple missions reveals where efforts are best focused for largest improvement in performance to achieve a SoS.

4.4 Run-Time SoS Formation

4.4.1 *Soft Systems Methodology SoS Guideline*

A soft systems methodology (SSM) was created for run-time SoS case studies. The soft systems methodology has proven to be a valuable process for defining problems and subsequent goals (Checkland, 1993, 2000). For this research, SSM is used to articulate the SoS problem space and provide a foundation for modeling and simulation. The SSM process is used for the run-time case study, an auto battle manager, as well as a trial on a U.S. health care system case study for future investigation. A description of the SoS SSM is found in Appendix B with a summary explanation of the methodology. For the case studies, a SoS SSM guideline is created as well as a SSM model in accordance to the created guideline.

4.4.2 *SoS Mechanism*

The planned or unplanned formation of a SoS from a collection of interdependent systems does so in response to a collaborative mechanism such as the common composeable patterns discussed in the previous section. The nature of the constituent systems that compose the SoS is a paradox by which they maintain an autonomous behavior and yet join the collective in collaboration. A social function acts as the SoS global utility function (U) that describes the SoS constitutive mechanism that considers the value of collaborative relationships, and stresses preferences for action rather than an action of the individual systems for design-time design for influence and run-time by self-organization. This decision dichotomy is an example of typical decision makers that must balance cost and risk against achieving goals in a satisficing environment.

The execution of cooperation in a satisficing process assumes that systems make rational choices and is the normative model. The decision maker will estimate utilities of a choice, calculate the value, and choose the option with the largest value. Rational choice assumes decision makers have all available information that will influence a decision as well as the rewards. Simon in (Herbert A. Simon, 1956) proposed a bounded rationality whereby obtaining maximum utility as well as an optimized decision is not possible due to insufficient information and an inability to compare and weigh information against available options.

For the case of a SoS, the holistic system does not exist nor can be possibly satisficing, discussed in-depth in Appendix F, or optimized if it were not for their interconnectivity. An SoS mechanism is achieved when a social function $f(\cdot)$ provides choices for the constituent systems to fulfill their expected utility (EU) in a holistic interaction. The social function in a SoS mechanism articulates the utility function that describes the holistic design-time and run-time environments represented as $U > \Sigma u_i$ where U is the utility of the SoS and u_i is the utility of each of the constituent systems. It is the basis by which to analyze decision theory on why and how systems choose to collaborate to satisfy their EU in a collaboration scenario. This analysis leads to designing collaborative system mechanisms explicitly in an implicit amorphous environment of autonomous systems. The collaborative SoS mechanism is created by establishing a structure by which the constituent systems have an incentive to behave according to social function rules.

The SoS mechanism describes the social function that enables systems to cooperate fulfilling their utility functions to achieve a common goal and discussed extensively in Appendix G. The environment may be unidirectional or bidirectional whereby the social function and mechanism design rules and policies are modified per changes in the behavior of the constituent systems. The mechanism consists of two aspects adapted from (Russell & Norvig, 2003):

1. Ontology for describing the set of allowable policies that the constituent systems can adopt, and
2. The resultant rule that determines payoffs to the constituent systems given the policy profile of the allowable policies.

In collaborative systems, the constituent systems are allowed to make binding agreements. It is assumed that the agreements have been made over a range of possible alternatives, which is the basis of bargaining theory and is not detrimental to them, called Pareto optimality. A SoS mechanism implementation creates the social function executing the allowable policies that imparts to utility theory the reasons for systems to satisfy their EU. The mechanism may be as simple as a voting scheme whereby systems vote if certain rules are satisfied and the winner is determined against another set of rules.

The basis of mechanism design is understanding utility theory that originated from the work of von Neumann and Morgenstern, which began the game-theoretic maximization of the EU (Archibald *et al.*, 2006; Von Neumann & Morgenstern, 1953). Each agent in the game-theoretic possesses a preference ordering of possible actions as a function that others may take. A major drawback of the game-theoretic von Neumann-Morgenstern utility theory is that it does not take into account social relationships that may exist between the systems. The possibilities of cooperation and conflict are not considered as in the case of the tragedy of the commons, only the maximization of the system's utility function or self-interest is a priority (Archibald *et al.*, 2006; Stirling, 2005; Stirling & Frost, 2005). Other examples are:

- Nash equilibrium makes the assumption that individuals of a game have a non-cooperative action which can execute independently of other players assuming no agreements are made and thus exhibits autonomy in a SoS context (J. Nash, 1951; J. F. Nash, 1950). Systems will not cooperate if in so doing will lower their utility than that guaranteed by their non-cooperation. This mechanism is prevalent in economic systems since strategies are expected to change if greater benefits can be obtained. A dominant strategy that is truthful is a Nash equilibrium, maximin, or Bayesian equilibrium (Kim & Roush, 1987).
- Vickrey-Clarke-Groves (VCG) mechanism whereby bids are secret and the winner pays second-highest bidder's offer. It is assumed that the highest bidder in not truthful and will not be able to pay. The VCG auction has the key property of incentive compatibility, which ensures that each participating bidder maximizes their payoff only by truthfully revealing their private information. This results in the needs of the individual are aligned with the needs of the many (Groves, 1973; Groves & Ledyard, 1977; Krishna & Perry, 1998).
- Hurwicz mechanism design theory laying the foundation of understanding the properties of optimal allocations of resources given individual preferences, incentives, and private information (Hurwicz, 1973; Hurwicz & Reiter, 2006).

In order to design a SoS mechanism to consider the value of collaborative relationships, the mechanism must establish a social function that stresses preferences for action rather than an action of the system. Each collaborating system has two natural preferences:

1. Maximize its goal to achieve the decision or
2. Conserve resources and avoid risk at the expense of achieving the goal.

This decision dichotomy is an example of typical decision makers that must balance cost and risk against achieving goals. It is rarely the case that a decision maker accepts a goal regardless of cost and risk as well as the opposite effect whereby goals are disregarded in consideration of saving resources and risk avoidance. This is achieved in a satisficing environment whereby holistic goals are achieved in an equilibrium solution space (Gorod *et al.*, 2009; Stirling, 2003) and summarized in Table 5 with respect to the SoS characteristics.

4.4.3 *Satisficing SoS Mechanism Design Social Function*

Group social functions are not a logical consequence of rational systems based on self-interest. Group behavior is not optimized when individual constituents are optimized in pursuit of maximizing their EU as is done in conventional von Neumann-Morgenstern game theory. Satisficing game theory permits the direct consolidation of group interests, and as such SoS goals, as well as individual interests; discussed at length in Appendices F and G. Satisficing replaces individual interests for group interest via holistic a model of inter-system relationships (Stirling, 2005; Stirling & Frost, 2005) creating collaboration. The SoS characteristics become evident in a group of systems with common goals in a satisficing manner of fulfilling their self-interest EU and the group's social function EU referred by Stirling as a private utility and social utility respectively (Stirling, 2003).

The group's social function that elicits the common preferences of the participating systems creating interdependence of collaboration comprises the totality of preference relationships and the sense of belonging. The system's preferences are conditional in meeting requirements of collaboration. System interdependence is established as linkages are established with coherent system EU preferences resulting in SoS capabilities. The mechanisms by which preferences are selected comprise the social function. The interdependence function, a sub function of the social function, comprised of the SoS characteristics of belonging, connectivity, and diversity are based on the participant's ability to select or reject preferences to satisfy its EU. An example is the U.S. health care system

Table 5. Neumann-Morgenstern and SoS Satisficing Social Function Characterization.

SoS Characteristic	Neumann-Morgenstern Game Theoretic Design Characterization of Egoism	SoS Satisficing Game Theoretic Design Characterization of Altruism
Autonomy	Constituent systems seek to maximize own utility function u_i through self-interest. Egoist behavior.	Autonomy exercised by constituent systems to fulfill the purpose of the SoS by selection or rejection of preferences maintaining an acceptable level. Tendency towards situational altruism modifying own preferences for common goals.
Belonging	Social function $f(\cdot)$ that maps the set of the constituent system type profiles θi the set of all global profiles θ in a predefined interdependent manner. Egoistic behavior.	Constituent systems choose to belong on a cost/benefits basis via selecting or rejecting preferences to cause greater fulfillment of their own purposes, and belief in the SoS supra purpose. The option space of selected preferences determines the social function. Altruistic behavior.
Connectivity	Constituent systems send messages whereby a rule set assigns an outcome.	Dynamically supplied by constituent systems with every possibility of myriad types of connections between constituent systems.
Diversity	Constituent systems utility functions u_i behave differently and expected utility (EU) is different with dissimilar payoffs.	Increased diversity in SoS capability achieved by released autonomy and open connectivity.
Emergence	Global utility function U revealed by preplanned cooperation of constituent systems' maximizing utility functions u_i.	Global utility function revealed through the selection of harmonious preference links between constituent systems.

whereby the system does not exist if it were not for the various systems that comprise it such as pharmaceutical companies, hospitals, insurance companies, physician offices, medical training organizations, and emergency medical response teams. The disparate organizations combine in a unique and adaptive way by which they all influence one another at various levels to be functional in their autonomy and diversity and interdependent in their emergent collaboration via connectivity and belonging. The participating systems of the health care system have released a portion of their autonomy and thus their optimization to collaborate. The decisions in choosing the available options for the participating health care elements define their satisficing set of preferences (Chao, 2008).

In the example of the six health care constituent systems, a satisficing game is defined as n systems where $X = \{X_1, X_2, \ldots, X_6\}$ each with a finite utility u_i. Any n-tuple $u = (u_1, u_2, \ldots, u_6)$ where $u_i \in U$ is a joint option. The Cartesian product of all system's trade-space options are $U = u_1 \times u_2 \times \cdots \times u_6$. The chosen options are based on conditional relationships that consist of two primary select ability functions denoted as $S_1, S_2, S_3, S_4, S_5, S_6$, or rejecting preferences denoted as $R_1, R_2, R_3, R_4, R_5, R_6$. The participating systems thus have two "selves" comprising twelve arguments negotiated in the option space. One self embraces collaboration to obtain some value or maintains total autonomy and rejects collaboration. Each system takes into account how its preferences are influenced by the actions of others (Stirling & Frost, 2005).

4.4.4 *Multicriteria Decision Making*

A means to incentivize collaboration within a SoS and to institute collaborative control is to prioritize constituent system behavior among the many preferences. A simple example is selecting a communications route for a multifunctional link cell phone that must establish a link based on lowest cost and choose routes delivering data in the shortest amount of time. It has choices a_x comprised of (land-based cell radio, traditional radio, satellite) and evaluation criteria Cx of (Cost, Arrival Time, Quality). The prioritization becomes a multicriteria decision analysis problem (MCDA) whereby discrete decision-making concerning a set of choices afforded by evaluating a set of criteria (Belton & Stewart, 2002; Hwang & Lin, 1987; Hwang & Yoon, 1981; Keeney & Raiffa, 1993; Yoon & Hwang, 1995). The application of MCDA here is a tool by which to evaluate and design the mechanisms of a SoS.

Yoon and Hwang in (Yoon & Hwang, 1995) describe a group of thirteen MCDA taxonomy methods based on salient feature information classified as ordinal, cardinal, pessimistic, optimistic, and whether the information concerned the environment or attributes. The satisficing system decision-making capacity is optimistic in selecting preferences in meeting its goals regardless of risk, cost, and resources or is pessimistic and rejects preferences in achieving cooperative goals and conserves resources, reduces cost and lowers risk. The MCDA process generalizes a prioritization approach by evaluating a finite number of actions versus criteria. An optimal solution to a MCDA problem is one that maximizes each of the objective functions simultaneously. It is not likely to obtain an optimal solution that satisfies all criteria due to problems of conflicting objectives and preferences. The outcome of applying preferences represents a compromise solution that reconciles all the criteria and is satisficing (Belton & Stewart, 2002; Hwang & Yoon, 1981; Yoon & Hwang, 1995). A satisficing solution is a set of preferences, which meets all the aspiration levels of each of the attributes. At a minimum, the final solution is reached by screening out the unacceptable solutions via lower priority preferences. It provides an objective analysis of alternatives against a goal and the supporting choices of criteria, alternatives, and relative weights via preferential trade-offs (Belton & Stewart, 2002).

Multiple preferences deemed acceptable or unacceptable via two types of satisficing methods described as conjunctive and disjunctive respectively (Yoon & Hwang, 1995). A conjunctive method is when all attributes meet or exceed a minimum threshold such as all student U.S. Naval Aviators must run an obstacle course in less than three minutes and must pass all basic aviation academics to proceed to flight school. In this method, all requirements are minimally met. How well a student aviator performs in running the obstacle course is not important, only their meeting all requirements is salient. In the disjunctive method, meeting only one requirement is acceptable from a list of requirements. For the Naval Aviator that does well in a course on flight dynamics can proceed to flight school even though the student cannot run the obstacle course in less than three minutes.

Consider a set of alternatives $A = (a_1, a_2, a_3, \ldots, a_n)$ and a set of actions $X = (x_1, x_2, x_3, \ldots, x_n)$ forming a matrix M that is a set of actions and alternatives resulting in chosen tasks (t).

$$M = \begin{matrix} a_1 \\ a_2 \\ a_3 \\ a_4 \\ \vdots \\ a_n \end{matrix} \begin{pmatrix} t_{11} & t_{12} & t_{13} & \cdots & t_{1n} \\ t_{21} & t_{22} & t_{23} & \cdots & t_{2n} \\ & & & & \\ & & & & \\ & & & & \\ t_{n1} & t_{n2} & t_{n3} & \cdots & t_{nm} \end{pmatrix}$$
$$\qquad\qquad x_1 \; x_2 \; x_3 \cdots x_m$$

The entries in matrix M indicate the degree of the interaction of criterion a_n is satisfied by alternative x_m. In simple MCDA process, this reduces to a single holistic criterion by aggregating all the elements of the matrix by weighting the arithmetic mean on the row of the matrix by $\sum_{i=1}^{n}(w_i)(t_i)$. This approach assumes that the criteria are all-independent and does not consider the interaction of the criteria. A weighting factor of each criterion as well as criteria on the subset of criteria is required for w_i.

Uncertainty plays an important part in the building of any model concerning the structure of the decision problem model and the environment of the subsequent actions. MCDA needs to incorporate the recognition of reducing uncertainty, albeit it is always present in some form. For MCDA, two forms of uncertainty categorized as internal and external regard problem structuring and the environmental consequences of an action respectively. Internal uncertainty does not know all the possible alternatives and the imprecision of known alternatives and their subsequent trade-off space. External uncertainty does not know the consequences of the choices made given the alternatives. An in depth discussion found in (Belton & Stewart, 2002; Russell & Norvig, 2003) provides the acuity necessary to understand the effects of uncertainty on MCDA and decisions.

4.4.5 *Analytical Network and Analytical Hierarchy Process*

An important task for decision-making includes the evaluation of alternatives and subsequent preferences. The Analytical Network and Analytical Hierarchy Processes (ANP/AHP), developed by Thomas Saaty (Saaty, 1980, 2005), are used to derive weights in multicriteria decisions such as in economics, construction, and decisions regarding environmental effects of human activity (Belton & Stewart, 2002; Saaty & Vargas, 2001). The ANP/AHP theory and process is represented in detail in (Belton & Stewart, 2002; Saaty, 2005, 2008; Saaty & Vargas, 1991, 2001) and is not discussed in-depth, only in its application of supporting the SoS mechanism.

The ANP/AHP process is a method to understand a complex system from a holistic perspective combining two processes of deductive systems analysis by examining the functions of the components versus the components themselves. It is an indirect weight elicitation technique using paired comparisons regarded as the most common and superior technique to measure decision choices across multiple choices (Buede, 2000; Lee & Kim, 2000). It considers the problem or goal holistically including constituent

Table 6. ANP/AHP Scale of Pairwise Comparisons Modified From (Belton & Stewart, 2002).

Intensity	Definition	Explanation
1	Equal importance	Equally preferred
3	Moderate importance	Weak preference
5	Strong importance	Strong preference
7	Very strong importance	Demonstrated preference
9	Extreme importance	Absolute preference
2,4,6,8	For comparisons between the above values	Interpolation of a compromised judgment
Reciprocals of Intensities	Activity i when compared to activity j, then j has the reciprocal value when compared with i	A comparison mandated by choosing the smaller unit as the unit to estimate the larger one as a multiple of that unit

systems' functions within their environment as a network and within a hierarchy. It is used here in an SoS case study to describe the actions of the SoS social function as a proxy method for systems' decision-making processes of choosing collaboration through belonging or autonomy out of self-interest.

Both ANP and AHP preferences are measured on a semantic nine-point scale for pairwise comparisons shown in Table 6. The comparisons are entered into a pairwise comparison matrix with respect to their relative importance to their control criterion. A relative ratio scale of measurement is derived from a pairwise comparison of the elements at each level in the hierarchy based on the influence of the element in the above level for AHP or in the connecting node for ANP. Local priority vectors derived for all comparisons and the local priorities are synthesized to derive the global priority used in making the final decision of a preference. The weights reflect the priority of the element. An Eigen value is determined to find the normalized vector of weights to promote the relative importance of the attributes at all levels of the hierarchy. The normalized weights of all the hierarchy levels combine to determine the unique normalized weights corresponding to the last level accomplishing the goal. A consistency index $\text{CI} = (\lambda_{\max} - n)/n - 1$ determines the consistency of the evaluations of numeral assessment of the pairwise comparisons by computing the largest Eigen value $\lambda_{\max}$ where n is the number of choices. The comparisons are satisfactory when the consistency ratio (CR), determined as a quotient of consistency index divided by a random consistency index (RI) is less than 0.10; $\text{CR} = \text{CI}/\text{RI}$.

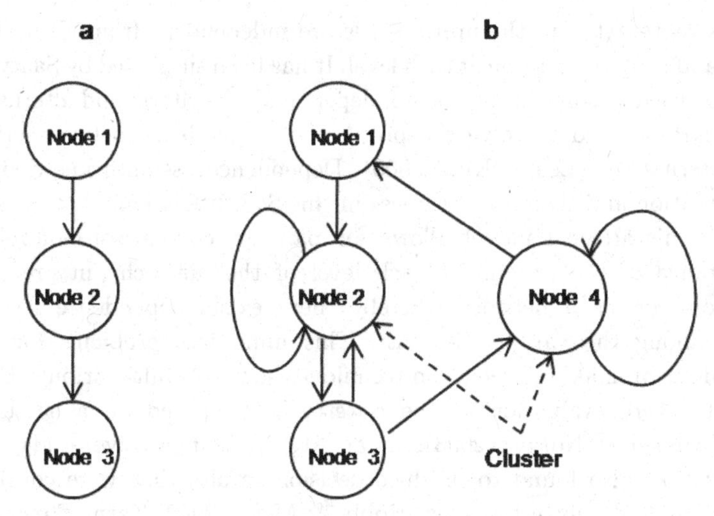

Fig. 10. Structural Difference between a Hierarchy-AHP and a Network-ANP: (a) Hierarchy; (b) Network.

The AHP represents a hierarchy approach and ANP represents a decision problem as a network of criteria and alternatives grouped as clusters shown in Figure 10. ANP is a generalization of AHP and consists of two parts: (1) a control network of criteria and sub criteria that control network feedback and (2) the network of influence that houses the factors of the problem and the logical groupings. Clusters represented by Nodes 2 and 4 of the Figure 10 Network (b) have elements of alternatives established for them and referenced as an inner dependence. The arc between Nodes 3 and 4 of Figure 10 — Network (b) expresses an outer dependence with respect to a common property. Thus, the alternatives within the node may influence other alternatives within the same node and those in other nodes with respect to each of several properties. Inner and outer dependencies are advantageous as they represent the concepts of influence between clusters and nodes (Saaty, 2005). A disadvantage of using ANP is structuring the problem with functional dependencies that allows for feedback among clusters in a network. In this scenario for SoS, there may not be feedback given a cluster's utility has not been met in which it decides to cooperate. The use of ANP as a SoS decision-making process proxy is for future research.

An advantage and value of using the AHP approach is the criteria used for the evaluation of the alternatives are arranged in a hierarchy representing their simultaneous interaction. Unlike ANP, it assumes the factors represented in the upper levels are independent from all its lower levels and from the criteria at each level. It has been suggested by Saaty that AHP be used to solve problems of independence of criteria and alternatives and ANP be used to solve the problem of dependence between criteria and alternatives (Lee & Kim, 2000). Dependence assumes knowledge of collaboration and thus is a weakness in this SoS discussion.

The hierarchical model allows for pairwise comparisons of relative importance of the criteria at each level of the hierarchy in preference selection and as a network whereby there exists dependence feedback loops among the various elements. This minimizes problems found in other decision-making evaluation techniques such as Multiattribute Utility Theory where evaluation of the criteria is compared on a single flat level (Russell & Norvig, 2003). In an MCDA comparative study, AHP method was also found to produce decision results that were consistent with other well known methods (Hobbs & Meier, 2000; Karni, Sanchez, & Tummala, 1990).

4.4.6 *Auto Battle Management Aids SoS Mechanism*

It is largely recognized that any organization must cooperate at various levels and true decentralization and autonomy cannot exist for the survival of the organization (K. J. Arrow, 1977). The example here is not about true decentralization without C2 but not true autonomy either. The SoS Auto Battle Management Aids is concerned with various levels of appropriate response to force-wide threats in a collaborative manner. Warfare area commanders initially establish the collaboration of battle group assets in the form of plans as a design-time SoS. This resembles a hierarchy decomposed from values and goals detailing a set of choices and alternatives based upon preferences made at each level of the hierarchy depicted in Figure 11. A top-down hierarchy beginning with a core value and goal is Value Focused Thinking (VFT) whereby a proactive approach based values serves to construct the decision process and creates the set of alternatives (Buchanan, Henig, & Henig, 1998; Corner, Buchanan, & Henig, 2001; Keeney, 1992). In reality, the alternatives are generally pre-determined from a set of preferences and in this example, MCDA is a process by which preferences are selected as a SoS design-time approach.

A battle group's plans are effective if the conflict and unfolding events are consistent with pre-conflict intelligence. However, given that most scenarios are indeterministic, the probability that events unfurl per

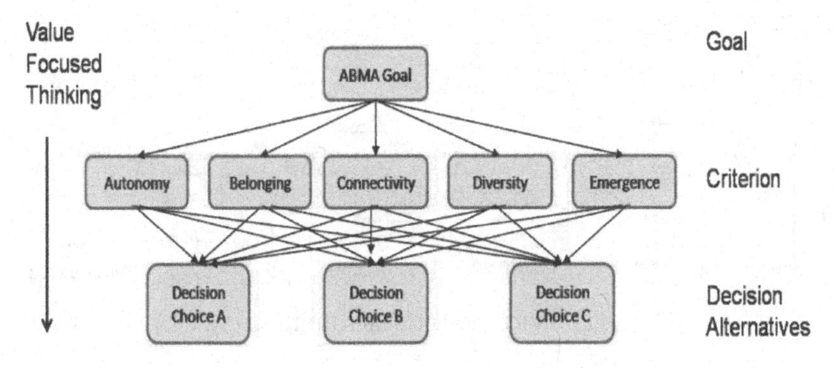

Fig. 11. Value Focused Thinking Hierarchy.

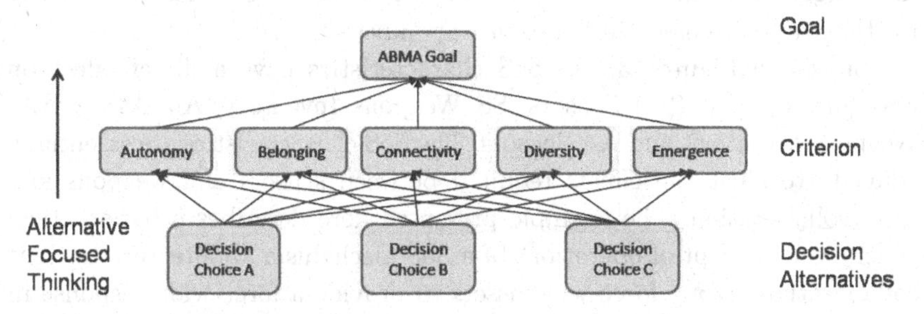

Fig. 12. Alternative Focused Thinking Hierarchy.

pre-determined plans are not likely. As the events unfurl, the plans become historical in nature and the deployment of assets and reaction of the warfare area commanders are based on training, experience, and communications. It is expected that many subgroups of the battle group have not trained with one another and operate with organizational warfare skills they know best as autonomous units. The situation and plans result in a "come as you are" preparedness necessitating a run-time SoS. The battle group warfare commanders react based upon alternatives available to meet higher-level goals shown in Figure 12 as Alternative Focused Thinking (AFT). The process of AFT does not afford creativity or innovation but is representative of most situations in real-time (Buchanan *et al.*, 1998; Corner *et al.*, 2001; Keeney, 1992). The MCDA process must consider mechanisms that allow criteria to consider new alternatives and a new sort of known alternatives (Corner *et al.*, 2001) and perform at the speed of battle.

Battle group decision alternatives are aligned with commander's intent through VFT hierarchy as design-time SoS and the alternatives align with commander's intent via AFT as run-time SoS in a circular fashion. The

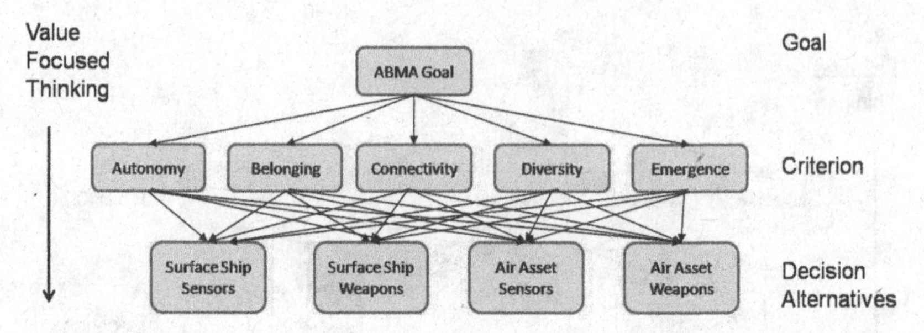

Fig. 13. VFT Battle Group AHP Hierarchy.

SoS ABMA mechanism macro architecture creates a pattern of responses to changes in the environment to elicit a force-wide response of the Navy Battle Group discussed in detail in Appendices F and G.

Shown in Figure 13, the SoS characteristics have a direct effect on how Surface Ship (SS) Sensors, SS Weapons Inventory, Air Asset (AA) Weapons Inventory, and AA Sensors. The SoS characteristics are essentially "dialed" to elicit the effects required between sensors and weapons and their collaboration. The example presented here describes a hypothetical generalized concept of operations of a SoS mechanism architecture to elicit collaboration among force-wide assets to provide a force-wide response in lieu of platform-centric responses to a force-wide threat. Platform-centric assets possess a SoS design-time preplanned rules to respond to air or surface threats. Such mechanisms could be Rules of Engagement, Tactics, Techniques and Procedures (TTP), or voting systems whereby preferences are established during design of the systems and their coordination and cooperation creating collaboration with other systems in a VFT process.

4.4.7 *Mechanism Scenario MCDA Model*

A scenario based SoS mechanism ABMA model is created to demonstrate the interactions of alternatives using the SoS characteristics as criteria. The social function established for the SoS characteristics' preferences (denoted as $\succ$) are Emergence $\succ$ Belonging $\succ$ Connectivity $\succ$ Diversity $\succ$ Autonomy to instrument the ABMA preferences in Appendix G and executed in Table 7. In this scenario, SoS capabilities via collaboration are a priority and thus the characteristic of emergence. In order to drive emergence, the characteristic of belonging is required because assets must impart their utility function and relinquish levels of autonomy. The characteristic of

Table 7. AHP ABMA SoS Characteristics Criteria.

ABMA Mission Collaboration	**Emergence**	**Diversity**	**Connectivity**	**Belonging**	**Autonomy**	**Priority**
Emergence	1.00	7.00	5.00	3.00	9.00	0.495
Diversity	0.14	1.00	0.20	0.20	3.00	0.064
Connectivity	0.20	5.00	1.00	0.33	5.00	0.152
Belonging	0.33	5.00	3.03	1.00	7.00	0.255
Autonomy	0.11	0.33	0.20	0.14	1.00	0.034
	Cl = 0.093		$\lambda_{max} = 5.37$			

connectivity is vital because there must be communications either direct or indirect. The characteristic of diversity is a lower preference because diversity of multimission platforms transforms themselves as a function of their utility functions and agreeing to collaborate. It is recognized that diversity provides SoS capabilities, but in this scenario; more of the same is a higher preference than autonomy.

The scenario to stress SoS emergent capabilities by relinquishing autonomy establishes the priorities of Table 7. The characteristics of emergence and belonging are dominant with the characteristics connectivity, diversity, and autonomy a good distance from the dominant characteristics. The dominant characteristics of emergence and belonging are the values to drive four choices of SS Sensors, SS Weapons Inventory utilization, AA Sensors, and AA Weapons utilization in Table 8. For emergence, the preference priorities are SS Sensors and AA Sensors because the ABMA must sense threats in order to adjudicate them. If the battle group assets cannot sense threats, the battle group cannot stop what it cannot see. In this case, the ABMA is looking for collaboration between surface based sensors and air based sensors. Typically, surface based sensors are limited to the horizon, but not the case for air assets. Through surface and air based sensor collaboration, the force of the battle group is extended beyond the horizon. For belonging, the priorities are similar, but SS Sensors and AA Sensors priorities are nearly equal which means their collaboration is a priority for this characteristic. AA Sensors slightly lead SS Sensors for the advantage of over-the-horizon battle group sensing. The two weapons priorities are reasonably close which means the two sets of sensors and weapons are collaboratively paired. From a holistic SoS ABMA perspective, the global effects place emphasis on ship sensors followed distantly by air sensors and weapons.

Table 8. SoS ABMA Sub Criteria Priorities and Alternatives.

	SS Sensors	SS Weapons Inventory	AA Sensors	AA Weapons Inventory	Priority
Emergence					
SS Sensors	1.00	3.00	4.00	5.00	0.534
SS Weapons Inventory	0.33	1.00	0.50	3.00	0.170
AA Sensors	0.25	2.00	1.00	3.00	0.219
AA Weapons Inventory	0.20	0.33	0.33	1.00	0.061
	Cl = 0.055		$\lambda_{max} = 4.16$		
Belonging					
SS Sensors	1.00	2.00	1.00	3.00	0.334
SS Weapons Inventory	0.50	1.00	0.30	2.00	0.170
AA Sensors	1.00	3.33	1.00	2.00	0.362
AA Weapons Inventory	0.33	0.50	0.50	1.00	0.124
	Cl = 0.042		$\lambda_{max} = 4.13$		

4.4.8 *MCDA SoS ABMA Simulation*

The MCDA SoS ABMA simulation experiment is discussed at length in Appendix G and shown in detail in Appendix H. The experiment uses scheme four, a hybrid social function of the four choices of an ABMA network summarized in Figure 14. The autonomous situation shown as scenario one of Figure 14 is the current day status quo whereby the battle group operates on local doctrine and preplanned responses. Each system is individually optimized assuming no collaboration with the attitude of "what is best for me is best for you." Scenario two is an ideal mode whereby all constituent systems of the battle group are decentralized but act as one in a force-wide response to threats. They respond to an amorphous doctrine in a satisficing behavior of "what is good for me is good for you." Optimization is force-wide versus individual ship platform optimization. This does not mean that ships of the battle group are sub-optimized to the level of losing a platform asset to a battle group threat. The scenario represents that the satisficing process has determined acceptable levels of risk to perform a force-wide response through collaboration. Scenario three is a representation of standard C2 conditions whereby autonomous systems follow instructions in responding to a threat. Optimization of ship platforms is predetermined for a force-wide response. There is no negotiation of the

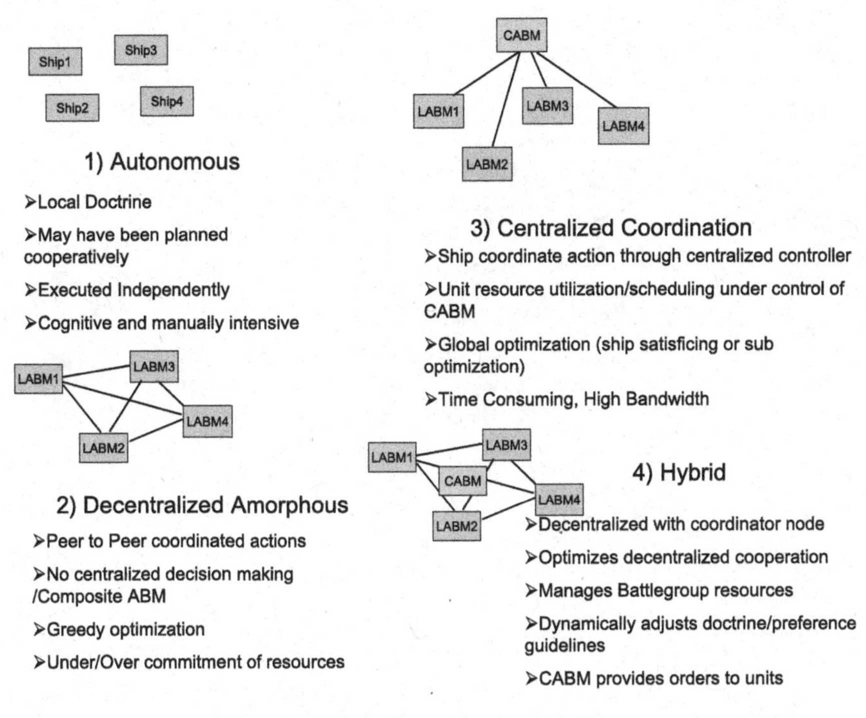

Fig. 14. Battle Group Cooperation Models.

risk trade space. Scenario four, an initial experimental step and focus of this research and case study to scenario two, is a compromise between C2 of scenario three and amorphous methodologies of scenario two. The scenario four chosen for the experiment is an evolutionary step from scenarios one and three. Evolutionary steps are more acceptable in DoD environments than revolutionary steps. The hybrid scenario of step 4 emphasizes cooperative C2 with sensors, weapons, sensor-weapons pairing, and communications.

Chapter 5

RESEARCH VALIDATION

5.1 Validation Overview

The simulation model tool used in this research experiment was the Lockheed Martin Systems Analysis Simulation (LMSAS), a proprietary model built by the Metron Oasis Group who leveraged its experience with the Naval Simulation System (NSS). This model includes representations for land attack missions, defense from cruise missiles, and improved command and control (Metron, 2008). The foundations of LMSAS are rooted in accredited modeling and simulation for the U.S. Navy. The modeling and simulation validation followed the guidelines outlined by the U.S. Navy to accurately represent as possible real world counterparts as directed by (Department of the Navy, 2004).

The approach chosen in this research is that of a case study, a low-constraint method, or qualitative research method. Higher constraint projects require the greatest amount of controls at all levels and rely on formal procedures to provide controls that increase the validity of research. Lower-constraint research validity depends on the clarity of thought (Graziano & Raulin, 2004). Low-constraint methods are a precursor to higher-constraint methods and questions to increase the confidence of causal action. Weaknesses in this method are:

- Poor representatives — only a small number of groups are studied for design-time and run-time being CAS and an ABMA respectively; cannot confidently generalize based on the observations and data reported in this research due to low level of controls by which cannot apply statistical analysis controls that are not part of the study;
- Poor replicability — the advantage of low-constraint research is also its liability, which is flexibility; different methods and examples

used following this research may produce different results; the implementation of the ABMA in this research used proprietary technologies such as data fusion engines shown in Appendix H and DoD sensitive data. In the case of CAS, identifying patterns was a manual process that is difficult to repeat.
- Tendency to over-interpret data and results; must avoid causal inference from the data.

The strength of low-constraint methods or of a case study:

- Provides for flexibility in the study of an unknown phenomenon and provides an exploratory basis for higher-constraint research; there has been very little study of the dynamics of a SoS as design-time and run-time and is exploratory in nature;
- Describe events not previously observed, identify contingencies, and negate general propositions of opposing observations.

The testability and validation of the SoS aspects of design-time and run-time are:

- Define autonomous systems and ability to choose;
- Define belonging and interoperation;
- Define changes in system utility or environment; and
- Define and measure results of belonging and interoperation.

5.2 SoS Design-Time

The design-time methodology presented in this work is based on the ability to derive patterns for composeable engineering and is truly exploratory. This execution of the CAS case study resulted in the unexpected discovery of patterns during the creation of the compositional process. By locating and evaluating patterns of disparate autonomous systems' interoperability, syntactic and semantic engineering became more easily performed. The identification of patterns using this process as compositional engineering has been met with some limited success by Lockheed Martin due to the laborious nature. Automation of pattern identification and evaluation is required to reduce error and improve productivity. This activity was trialed on a Marine Air Group Task Force (MAGTF) problem involving Digital CAS. The concept and process has been found to bear merit with the Marine Corp as well as other research activities whereby disparate legacy systems dominate. Demonstrated observables are:

- Reduction and composition epochs demonstrated via diachrony,
- ZF framework of SoS characteristics of connectivity demonstrated through the aspects of What, Who and, How.
- Common patterns in SoS composition leading to SoS Design for Influence identified and thus defining the social function, or SoS utility function, for the particular mission thread,
- Design for Influence methodology demonstrated whereby systems can be more readily designed to interoperate versus a single monolithic system.

5.3 SoS Run-Time

The experimental simulation of an SoS ABMA consisted of fifty Monte Carlo runs discussed in greater detail in Appendices G and H. The experiment was established as an AFT event whereby autonomous system alternatives negotiate for collaboration at a higher level, adjudicated against a social function shown in Figure 15 fulfilling commander's intent. The social function in this experiment was to neutralize the threat, a raid of cruise missiles, in a cooperative scheme reducing the amount of inventory expended, losses of battle group assets, and do so through minimization of risk for autonomous systems.

The measures of effectiveness used in the run-time SoS experiment are the results of SoS ABMA behavior that is negotiated among the

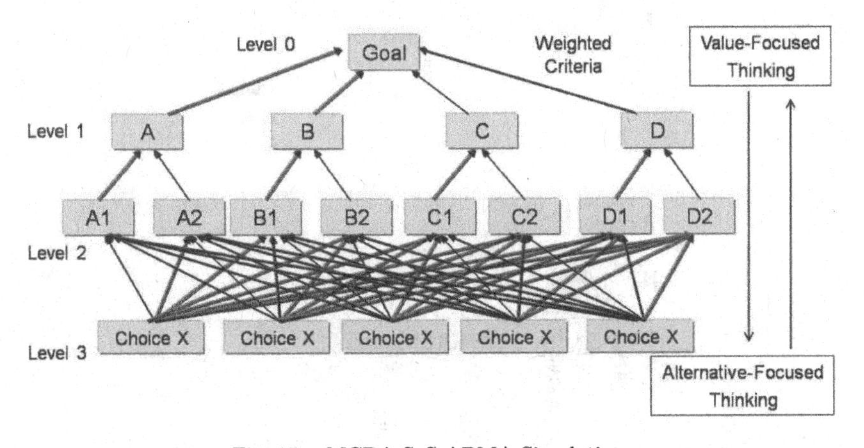

Fig. 15. MCDA SoS ABMA Simulation.

autonomous battle group systems and reported visual relative comparison due to DoD sensitivities:

- Battle Group Hits — the number of undefended impacts by enemy cruise missiles;
- Battle Group Weapons Expended — the measurement of two types of defensive missiles, one that is guided to the horizon and one that is not;
- Decision Time — the amount of time it takes to engage an incoming threat;
- Range — the distance required to attrit a threat.

The results of the experiment, discussed in detail in Appendix G and implementation details shown in Appendix H, reveal positive results of collaboration social function schemes for SoS constituent autonomous systems. In the case of an ABMA Naval battle group countering a cruise missile raid, there were fewer battle group losses from a baseline of no ABMA collaboration, to ABMA collaboration of surface assets as ABMA1, to additional air assets of ABMA2 shown in Figure 16. Through collaboration of resources, mitigating autonomy for belonging and balancing risk of belonging to share resources, damage to the battle

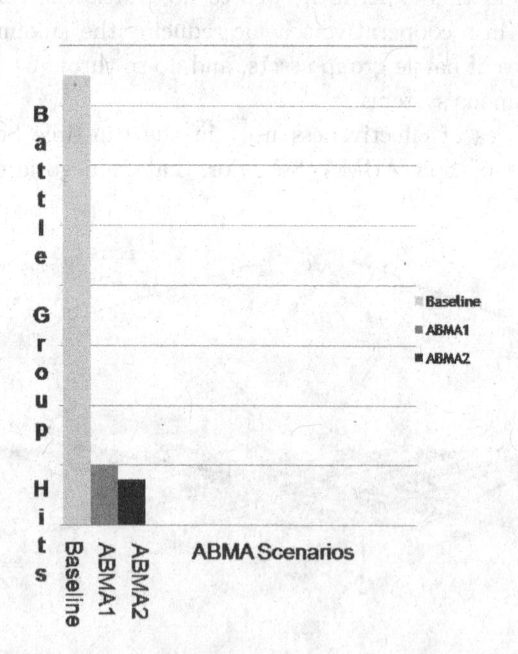

Fig. 16. Relative Comparison of Battle Group Enemy Strikes.

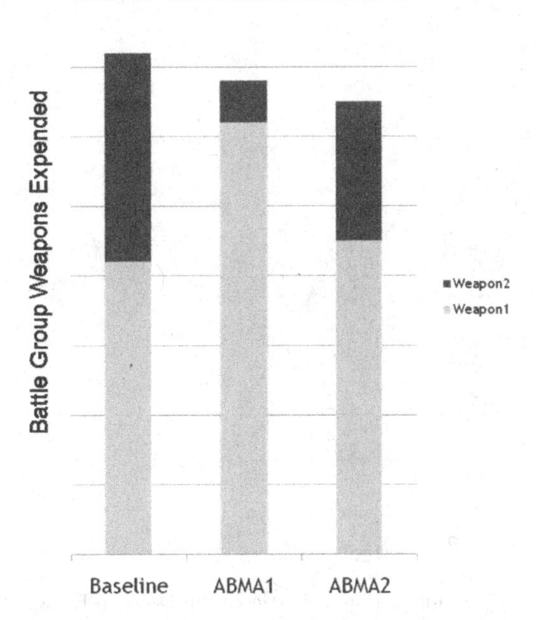

Fig. 17. Relative Comparison of Battle Group Weapons Expended.

group is reduced. Actual axis scales are not displayed for reasons of DoD sensitivities.

In concert with less battle group strikes from air threats, the expenditure of two types of counter air threat weapons were better managed shown in Figure 17. In the baseline of Figure 17, there is autonomous treatment of weapons and all assets respond accordingly to air threats. There is an autonomous response of weapons use but not a force-wide response. The ABMA1 data shows a smarter use of weapons via better resource management force-wide. With the addition of air assets in ABMA2, there is a change in weapons selection noted as Weapon2, which is a missile of greater distance and capability. Weapon2 is sparingly used in ABMA1 versus the baseline and ABMA2. In the baseline, Weapon1 and Weapon2 are used extensively with the greatest number of enemy strikes ensuing. With a smarter use of weapons resource deployment shown in ABMA1, there is a force-wide social function cognizant decision to use what weapon against which threat, by whom, and when. The ABMA2 scenario further reinforces this with the addition of air assets whereby Weapon2 is used more extensively with less inventory expended overall with fewest enemy battle group strikes.

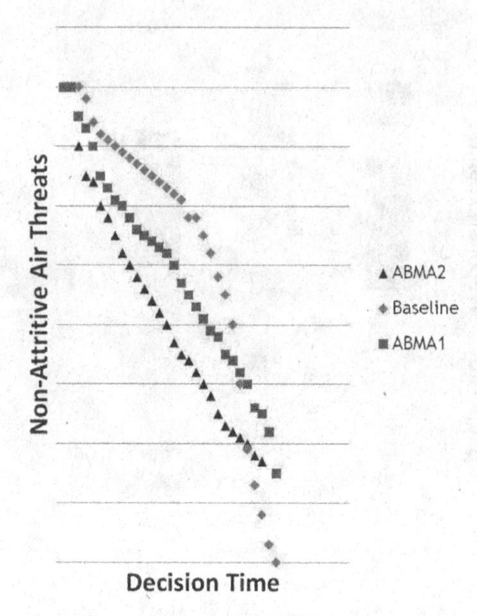

Fig. 18. Relative Comparison of Battle Group Decision Engagement Time.

Decision time is reduced as the battle group responds in a force-wide method versus all battle group assets responding autonomously. Shown in Figure 18, response time against incoming threats is reduced. The baseline curve intersects the x-axis as not all threats are attrited as decision time expired. The prosecution of threats is faster in ABMA1 and ABMA2. Better decisions of what threat to shoot with what and by whom is performed faster resulting in less weapons inventory expended and less battle group losses.

In the context of engagement range shown in Figure 19, threats are attrited further from the battle group, thus the greater use of the longer range Weapon2 of Figure 17. The curve of the ABMA scenario of Figure 19 is flat at the beginning as threats are seen but not engaged. The trend towards Quad D is the goal of attriting threats sooner and at greater range.

This experiment demonstrated SoS social function that provided a foundation for autonomous systems to adjudicate their alternatives of maximizing their utility functions against a SoS utility function, managing their autonomy against SoS belonging, to elicit a force-wide response to a force-wide problem in a synchronic process whereby their dynamics were within their synchrony boundaries. In the scenario baseline, optimization was towards a single mission that detracted from other missions. The

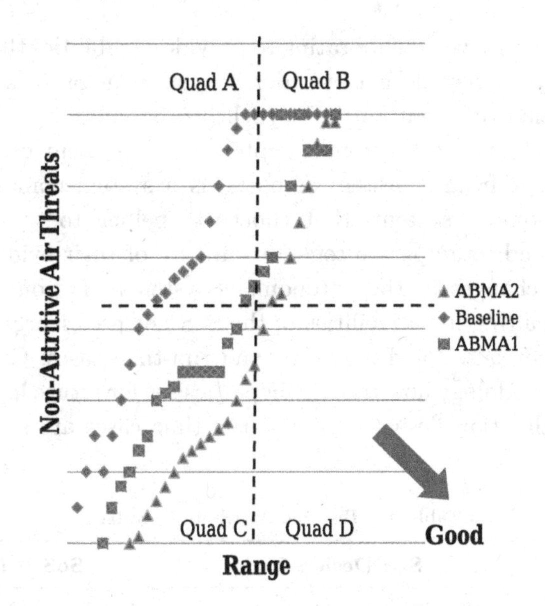

Fig. 19. Relative Comparison of Battle Group Threat Engagement Range.

force-wide response realized the dynamic approach to satisficing force-wide needs. The 'good enough" solution was acceptable and thus a level of sub-optimization was acceptable per real-time preferences of position and sensor and weapons availability.

5.4 Validation Summary

The research approach discussed in Chapter 3 was demonstrated for design-time and run-time SoS whereby constituent systems balance their utility of autonomy and belonging. Through their collaboration, emergent capabilities result and act as feedback to autonomy and diversity in their preferences in real-time SoS. For design-time SoS, their syntactic and semantic patterns provide in SoS ZF E^3 a social function whereby the SoS forms a mechanism of collaboration through identified interoperability. The systems transform through a diachronic process.

For the run-time case, the SoS social function adjudicated the needs of the many versus the needs of the few. The run-time SoS alternatives as "come as you are" collaborated to defeat a force-wide threat. The neutralize threat social function using satisficing by managing autonomy versus the risk of belonging was demonstrated, fulfilling their autonomous utility functions and yet collaborating to provide capabilities that individual autonomous systems could not provide via synchronic process of tolerating dynamic system events within their synchrony boundaries.

The hypothesis stated from Chapter 1, "... system of systems (an SoS), as distinct from a system of parts, is a system comprised of pre-existing autonomous systems that choose to belong to the dynamically forming SoS, and interoperate together in spite of their evident diversity, is a result of changes in the autonomous systems' environment or their own utility resulting in capabilities of the SoS not presciently designed" is shown to be not false for design-time and run-time cases. Given the low-constraint methodology how true the hypothesis is for future high constraint testability. Validation design-time and run-time cases are summarized in Table 9.

Table 9. Testable Attribute Results.

Testable	SoS Design-Time	SoS Run-Time
Autonomous systems and ability to choose	Chosen by users to fulfill a need; ZF E^3 demonstrated in SoS connectivity of patterns	Self-organized per preference alternatives adjudicating to a social function per satisficing; balancing autonomy and belonging via own utility vs. risk
Belonging and interoperation	Diversity of functions connecting via similar patterns via composition; users defined belonging to enable interoperation due to foreseen value	Communicate preferences of own utility adjudicating against social function; interoperated per satisficing risk of autonomy
Changes in utility or environment	Diachronic changes in systems allow connectivity due to dynamics extending outside synchronic boundary; SoS utility foreseen by users and provide governance per epoch and SoS mission thread	Synchronic acceptable dynamics causing changes in own system utility enabling SoS capabilities of collaboration
Results measurement	Designed and integrated connectivity offers belonging inducing emergence per design for influence per ZF perspectives and aspects; new communication paths using common patterns enable dynamic mission threads	Needs of many met through satisficed autonomous collaboration; increase of belonging with lowering of autonomy while meeting acceptable self-interest utility (u) increases capability, reducing resources

Chapter 6

CONCLUSIONS AND FUTURE RESEARCH

6.1 SoS Collaborative Formation

The machinery of the SoS formation as contrasted to "regular systems" is summarized in Table 10. This research explored the concepts of a social function defining the mechanism of how a SoS forms and constituent systems collaborate. In a design-time SoS, patterns defined the SoSI based on ZF E^3 user perspectives who defined the value for a SoS. A diachronic process provided the execution of the ZF E^3 whereby through SoSI, constituent systems exceeded their synchrony and evolved to provide new capabilities to implement a SoS. If it were not for the temporal aspect, the constituent systems could remain in their synchrony affecting a SoS similar to run-time SoS. In the run-time SoS, synchronic change allowed their dynamics to collaborate and remain risk tolerant to affect belonging through choosing their alternatives in an AFT process. This is summarized in Figure 20.

6.2 Future Research

6.2.1 *SoS Social Function*

Further experimentation and concept validation is required in high-constraint approaches for design-time and run-time SoS. For design-time, a full implementation of ZF E^3 methodology with stakeholders of legacy systems is required. For run-time SoS, a non-DoD application is required to test social function mechanism concept as the reason for SoS collaborative formation. An approach application of design-time as well as run-time considerations is applying the concepts presented in this research to the U.S health care enterprise and shown as an SSM model in Appendix B.

Table 10. SoS Formation Conceptagon.

Conceptagon	"Regular Systems"	SoS
Function structure process	Predefined autonomous systems and defined interfaces to predefined systems in a predicted environment; Hierarchical objectives and views	Composing autonomous constituents; Flat; Distributed; Systems decide based on preferences to satisfice utility function; Hierarchical internally and flat externally.
Relationships wholes parts	Predefined parts as hierarchical residue of appointed parents; Reductionist hierarchical engineering; Utility functions predetermined; Interfaces designed for predetermined external systems with predetermined static relations	Loosely predefined to undefined; Composeable; AFT methodology; Systems determine utility function levels; Determined based on immediate need to collaborate and satisfice utility. Wholes and parts are transitory. Constituent system remains intact.
Emergence hierarchy openness	Emergence due to unintended consequences of unintended system or environmental behavior; Not adaptively open, but dynamic per design. Rigidly hierarchical in design and operations.	Unpredicted consequences of systems satisficing utility functions; Evolution of constituent systems due to synchronic influence and design for influence. Hierarchy internal and flat external. Open systems except may choose to be closed (autonomous) to max utility function.
Communication command control	Central to designed decentral predicted and designed protocols via syntactic engineering with prescribed semantics; React to predefined commands with predefined control via sets of rules.	Central to decentral; predicted and designed omnipresent protocols; Amorphous command and control meeting SoS social utility establishing priorities of system preferences for constituents to respond.
Boundary interior exterior	Tight boundary. Predefined interior and predefined exterior of operations and interfaces.	Dynamic to adaptive boundaries whereby interior may change at design-time in compositional engineering; exterior will change due to satisficing of utility function and ability to open boundaries. True for physical and evolved systems.

(*Continued*)

Table 10. (*Continued*)

Conceptagon	"Regular Systems"	SoS
Transformations inputs outputs	Planned and designed transformations, input and outputs. Changes due to system owner in response to stakeholders.	Transformation due to synchronic and diachronic influence, variable inputs and outputs to meet SoS social function. Constituent systems will vary and be part of the whole that meets social function.
Harmony variety parsimony	System utility function defined as well as responses to syntactic and semantic communications per designed constraints.	Social function utility is king and will embrace all constituents that relieve autonomy for belonging. Constituents satisfice their own utility to belong. Social function utility assumes lowest cost for constituents to belong.

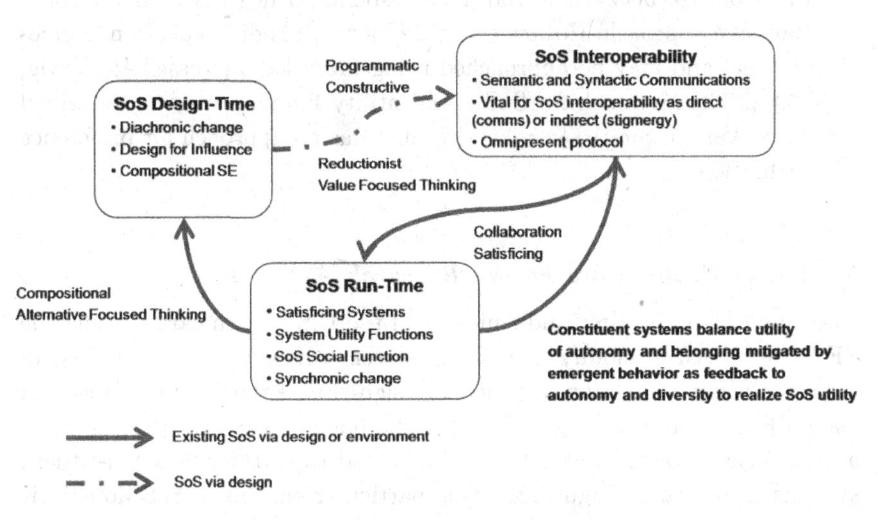

Fig. 20. SoS Collaborative Machinery.

In regard to mechanism of the SoS social function, future research topics are:

- The Technique for Order Preference by Similarity to Ideal Solution (TOPSIS) is based on the idea that the chosen alternative shall have the shortest distance from the ideal solution and the farthest distance from the negative-ideal solution (Hwang & Yoon, 1981). This method has been shown to provide preference order solutions and is similar to the Hurwicz rule because it is an amalgamation of optimistic and pessimistic decision methods (Hey, 1979). This may solve the problem of non-normalized data as is required for the Hurwicz rule. The comparative analysis for a SoS is necessary.

- Fuzzy logic approach to MCDA in the SoS social function is an important approach to determine system preferences for collaboration as the preferences become more complex and humanistic. These systems and their description of interoperability become nebulous and incomparable. Nebulous goals and constraints result in nebulous or fuzzy attributes. There are techniques described in the literature that transform fuzzy attributes into discrete identifiers, but are not well suited for system choices of collaboration (S.-J. Chen, Hwang, & Hwang, 1992). The complexity of the interoperability of systems leads to greater nebulous reasons of belonging and eventual collaboration. Probabilities between 0 and 1 correspond to degrees of belief for a Bayesian approach. However, degrees of truth is different than degrees of belief and may be approached using fuzzy logic (Russell & Norvig, 2003). The approach for SoS social utility function may be described from (Verma, Smith, & Fabrycky, 2000) as an approach for preference evaluation.

6.2.2 *SoS Health Care Future Research Application*

The union of design-time and run-time of the U.S. health care is using the ZF E^3 approach to modify an existing system whereby utility functions of the constituent systems are satisficed. Design-time is the portion of redesign using ZF E^3 and satisficing is the run-time portion whereby preferences to an established social function are adjudicated to satisfice SoS constituent systems displayed in Figure 21. This particular enterprise SoS domain is particularly challenging:

- There is no central authority thus amorphous in behavior;
- Decision of one constituent system affects the options available to the others;
- Reformers can only improve local systems and thus no systemic change except for government policy; and

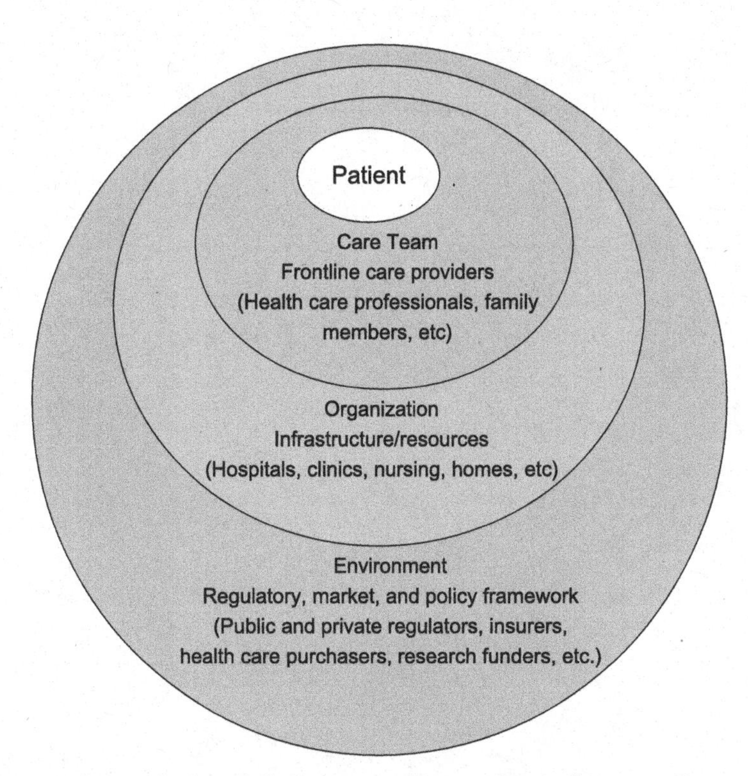

Fig. 21. U.S. Health Care SoS Constituent Systems (Reid, Compton, Grossman, & Fanjiang, 2005).

- Centralized approach or centralized medicine has not been found to be feasible.

The challenge is designing decentralized models of operation that render operation the overall system or health care SoS as effective as possible, very similar to classic industrial supply chain. The disparate functional entities of the SoS health care must collaborate for overall system performance. The solution space may be the intersection of conventional system engineering and economics using the concepts of this research as a methodology.

Appendix A

DESIGN-TIME COMPOSITIONAL SYSTEMS OF SYSTEMS ENGINEERING

The compositional process described in Chapter 4 is presented here in detail and shown in Figures 22 through 28.

Step 1: Reductional — C2 System functional allocation across all mission threads is populated in a database. The data is per epoch per mission thread decomposing to warfighting functions. The C2 systems are decomposed to applications, enterprise services, network services, and bandwidth.

Step 2: Reductional — Builds mission threads via use case diagrams, use case mapping and UML activity diagrams. The use case diagram depicts the actual systems CAS interfaces with such assets as Nail 51 which is a F-16 aircraft and NFCS which is Naval Gun Fire Support. The use case mapping provides further decomposition of team, objects, and actors of the actual mission threads through the constituent systems depicting the interoperability and required functionality. The UML activity diagram provides further decomposition of constituent systems in C2 functions of detect, control, and engage activities.

Step 3: Composition — Upon collecting the necessary mission thread data to portray the interoperability between the constituent systems of CAS, the mission thread is further evaluated by combining the data of Steps 1 and 2 to identify common patterns.

Step 4: Composition — subsequent syntactic and semantic connectivity supporting the C2 functionality of detect, control, and engage. Transmit, response, requests, command, and interrogation and response connectivity is identified via the common patterns among the constituent systems. Depict ZF Who, What, and How.

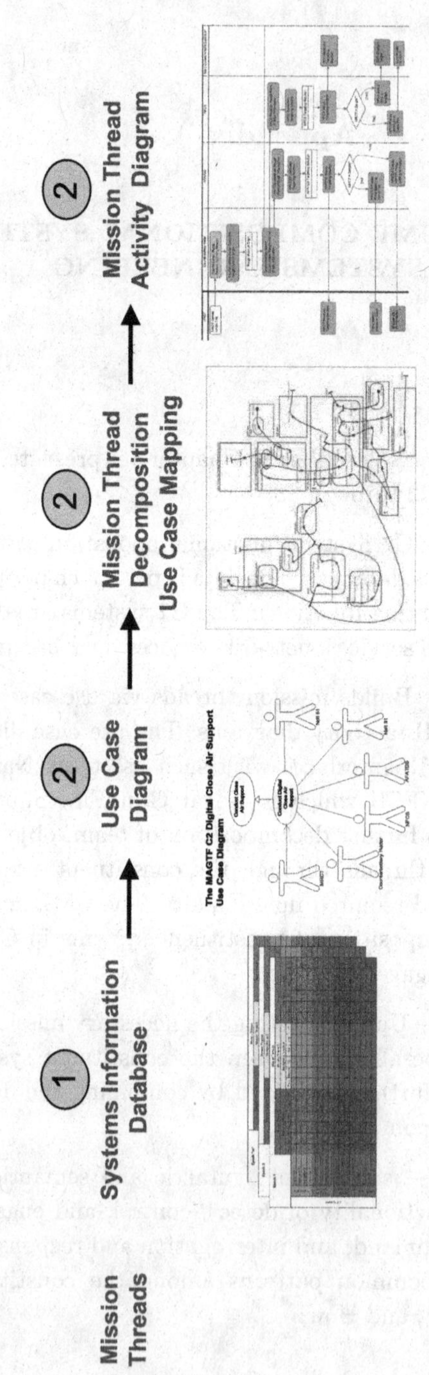

Fig. 22. CAS Decomposition.

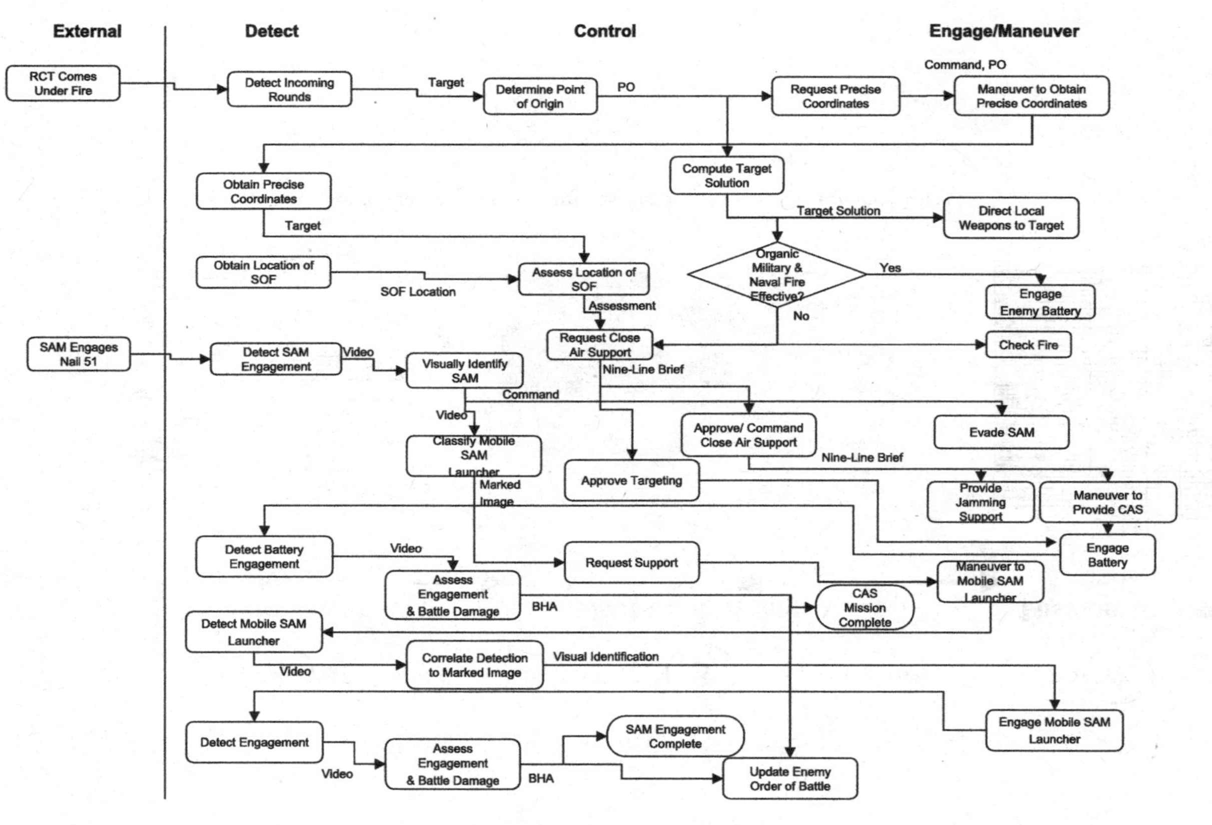

Fig. 23. CAS Composition Mini Thread Evaluation.

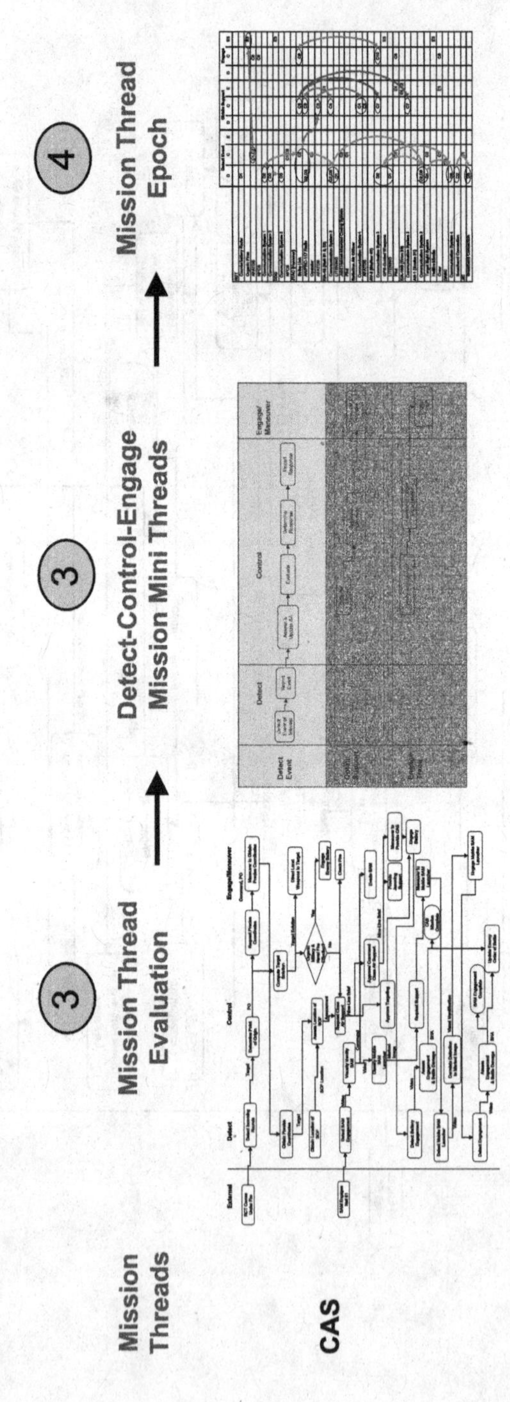

Fig. 24. Consolidated Mission Threads Per Epoch and Governance View.

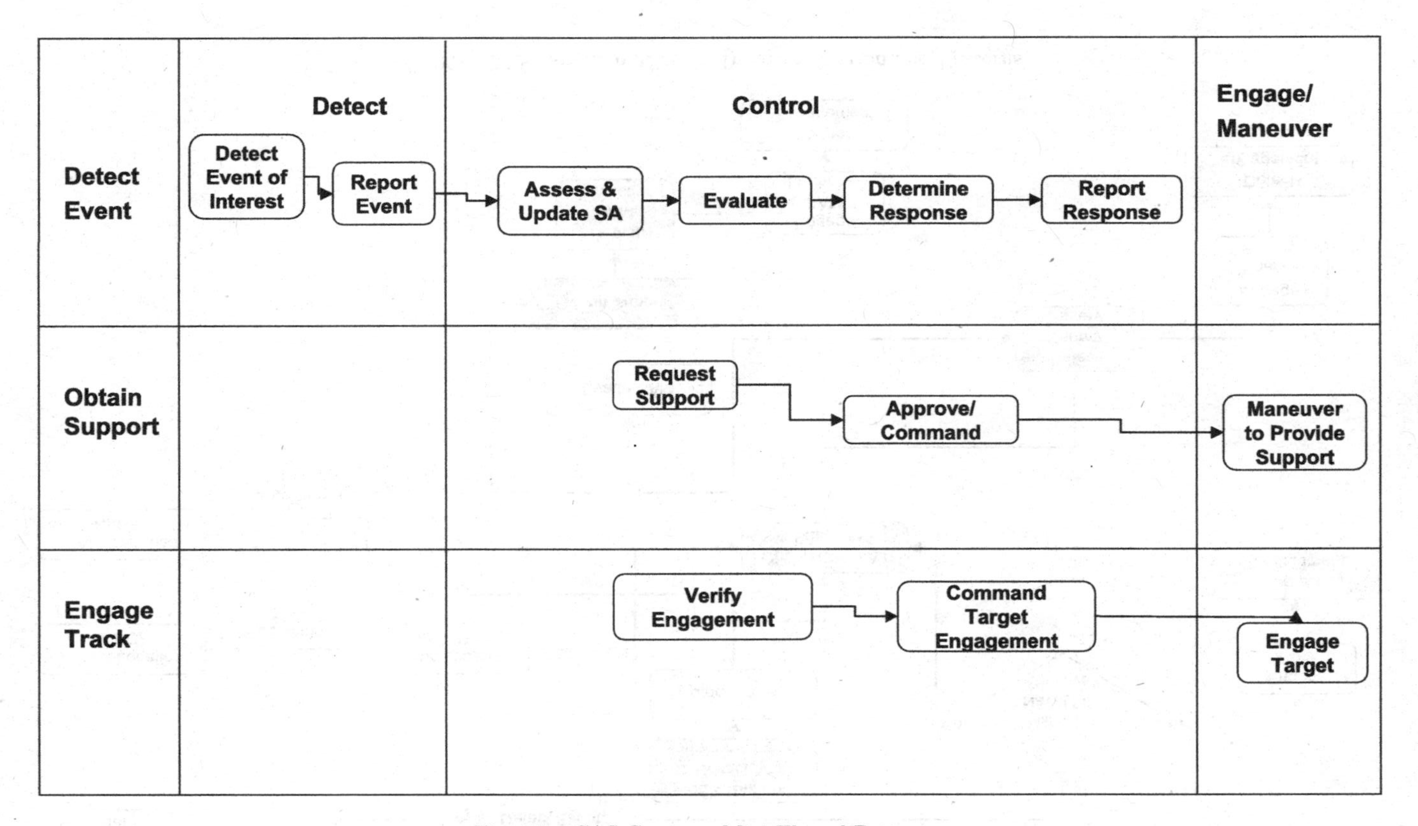

Fig. 25. CAS Common Mini Thread Patterns.

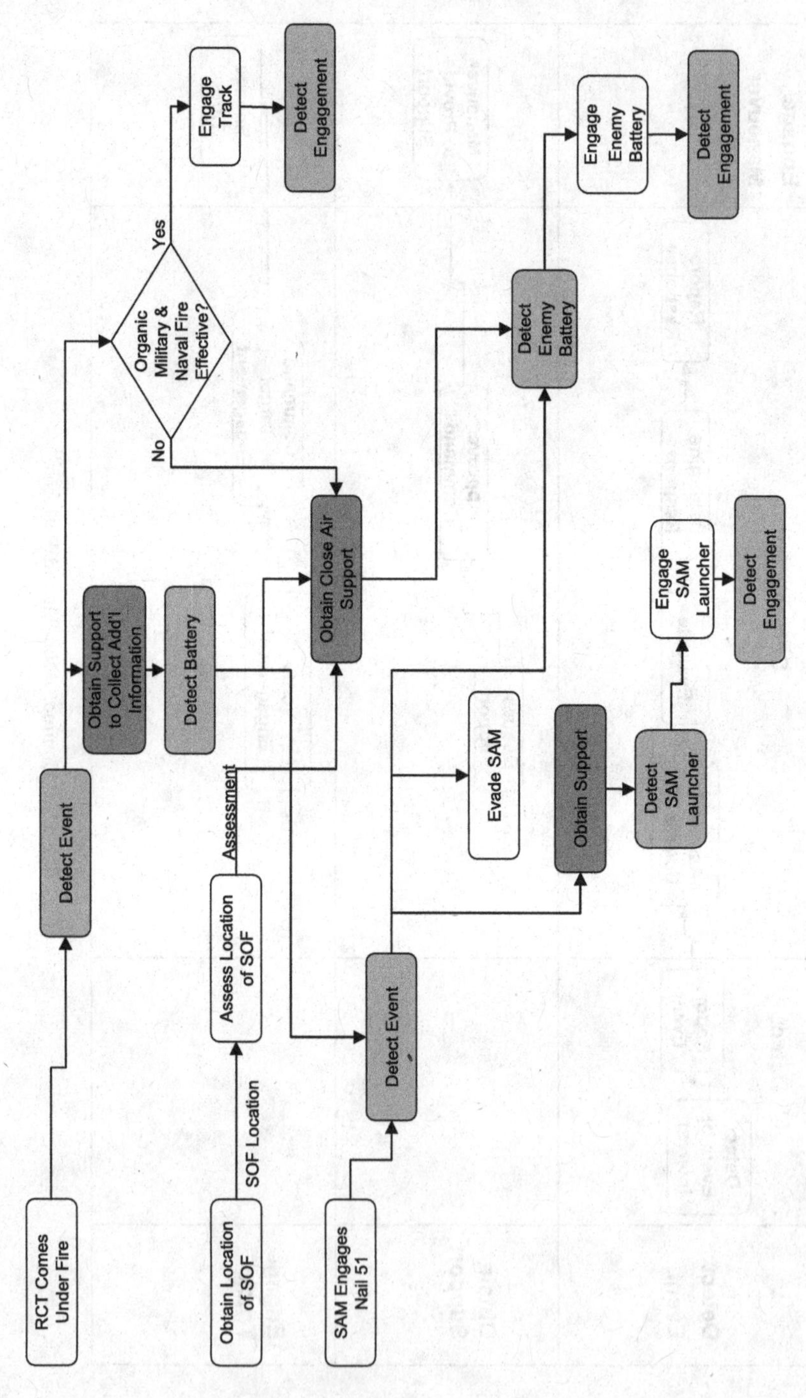

Fig. 26. CAS Common Mission Tread Evaluation and Patterns.

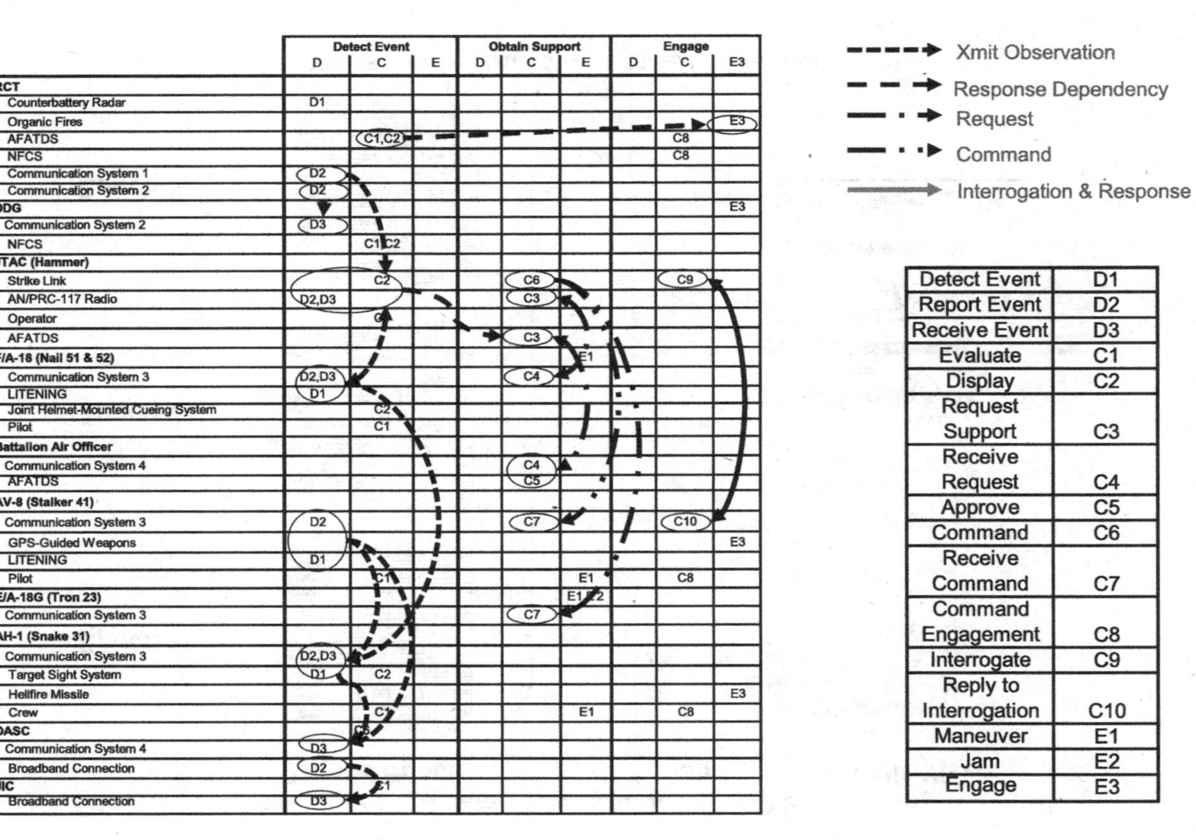

Detect Event	D1
Report Event	D2
Receive Event	D3
Evaluate	C1
Display	C2
Request Support	C3
Receive Request	C4
Approve	C5
Command	C6
Receive Command	C7
Command Engagement	C8
Interrogate	C9
Reply to Interrogation	C10
Maneuver	E1
Jam	E2
Engage	E3

Fig. 27. Common Patterns Define Mission Thread.

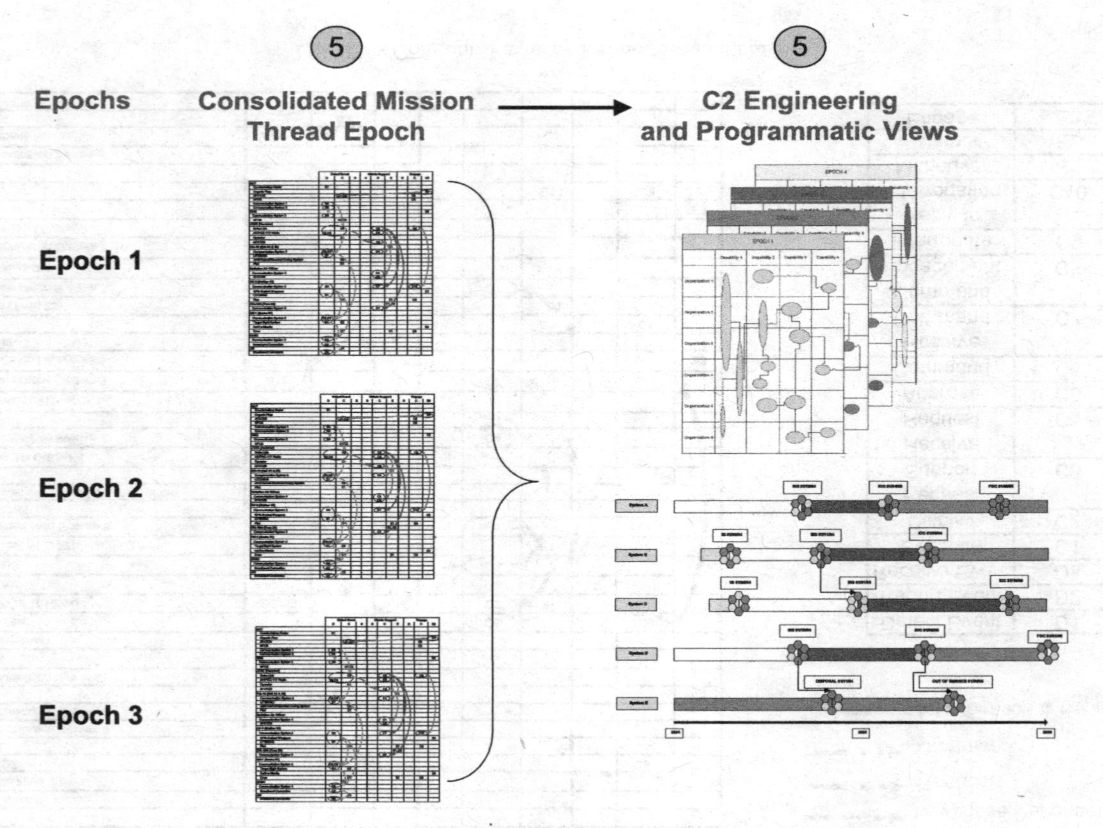

Fig. 28. Consolidated Mission Threads Interoperability.

Step 5: Composition — Connectivity an interoperability patterns are performed per epoch and mapped to programmatic views whereby the SoS is managed by interoperability capability and mission thread versus individual constituent systems.

Appendix B

SOFT SYSTEMS METHODOLOGY SOS GUIDELINE

For Figure 29, CATWOE is a mnemonic checklist for a problem or goal definition applied to the system that contains the problem, issue, or solution, rather than to the problem or goal itself. The "customers" of the system means those who perceive value of whatever it is that the system does. The "actors" meaning those who would actually carry out the activities envisaged in the notional system being defined. The "transformation process" is what does the system do to the inputs to convert them into the outputs. The "world view" provides the foundation for the Root Definition. The "owner(s)" are those who possess governance over the system. The "environmental constraints" include but are not limited to regulations, financial constraints, resource limitations, limits set by terms of reference, and so on.

Another two attributes of Root Definitions are the use of P-Q-R structure of the definition, and the focus on the 3Es principles when considering the transformation itself. This structure follows the form Do P by Q to contribute to achieving R. P answers the question What to do?, Q answers How to do it? and R answers Why do it?. The 3Es are used as a measure of performance of the system, typically Efficacy, Efficiency and Effectiveness.

The stages represented by the SoS SSM of Figure 30 supports the MCDA and SoS social function approach used in this research:

1. Enter situation considered problematical.
2. Express the problem situation.
3. Formulate root definitions of relevant systems of purposeful activity.
4. Build conceptual models of the systems named in the root definitions.
5. Comparing models with real world situations.

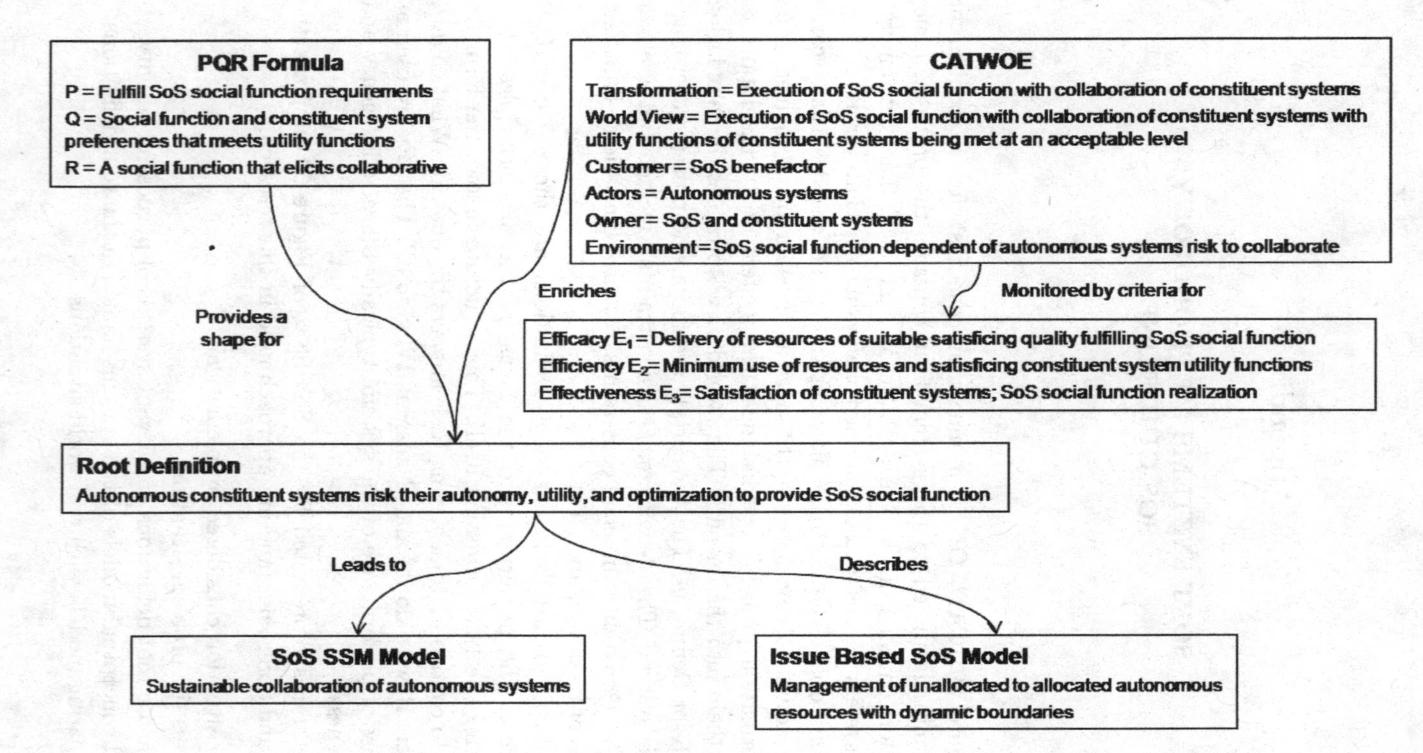

Fig. 29. Soft Systems Methodology SoS Guideline.

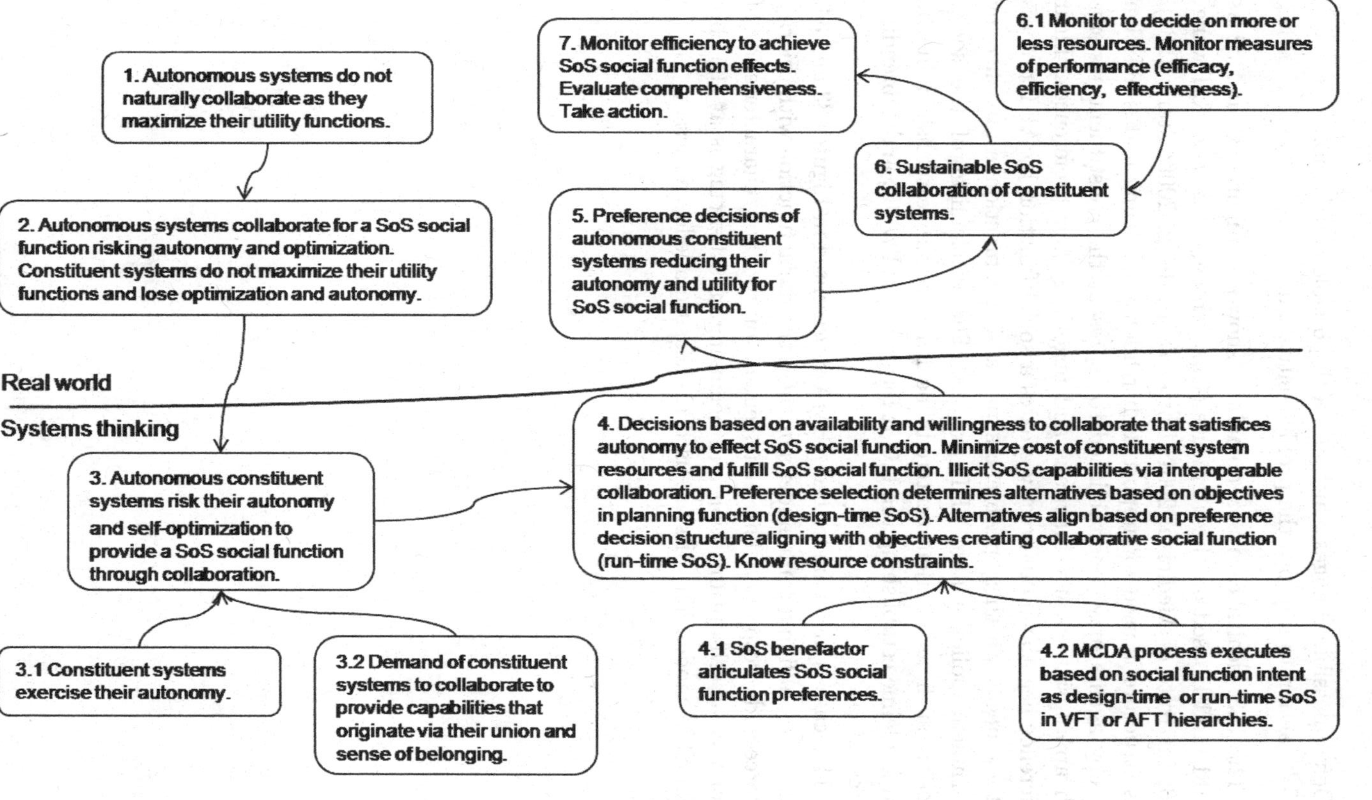

Fig. 30. Soft Systems Methodology Generic SoS Model.

6. Define possible changes which are both possible and feasible.
7. Take action to improve the problem situation.

The U.S. health care SoS domain is shown in Figures 31 and 32 as a SSM model based on the literature (Christensen, Grossman, & Hwang, 2009; Institute of Medicine, 2001; Porter & Teisberg, 2006; Starr, 1982). This structure is used to support higher fidelity modeling and simulation whereby agents may be established to represent the constituent systems with appropriate utility functions and negotiate their preferences. Their preference negotiation indirectly creates a social function by AFT process or they negotiate their preferences per a social function established by government policy via a VFT process. The mechanism of the social function such as "patient-centered" health care are established in the SSM guideline establishing the P,Q,R formula, CATWOE, and subtending process steps.

This case is also true for the ABMA represented in Figures 33 and 34. In this case, fulfilling commander's intent is the social function with priority of force-wide response. The management of the SoS characteristics of autonomy as self-optimization and belonging as satisficing is at the core of the SSM model. The 3Es are used as measure of effectiveness.

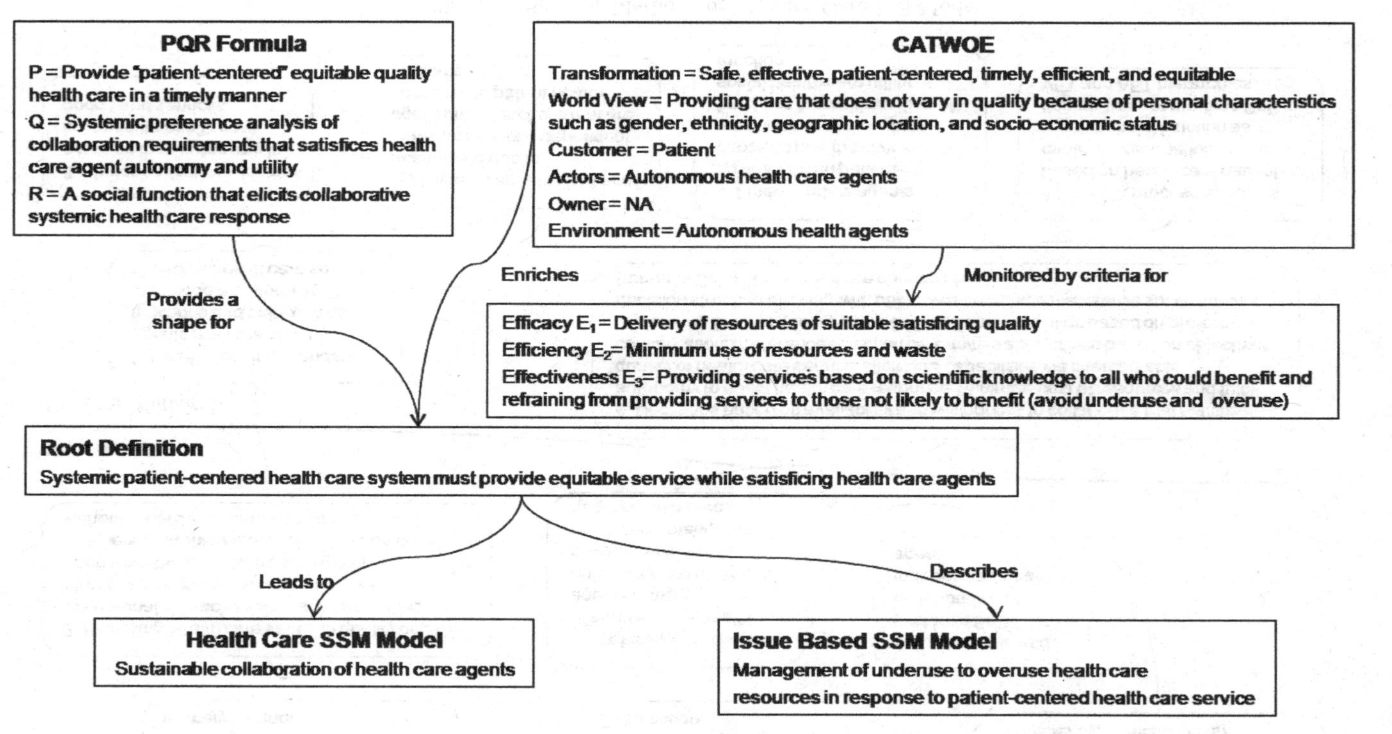

Fig. 31. Soft Systems Methodology SoS Health Care Guideline.

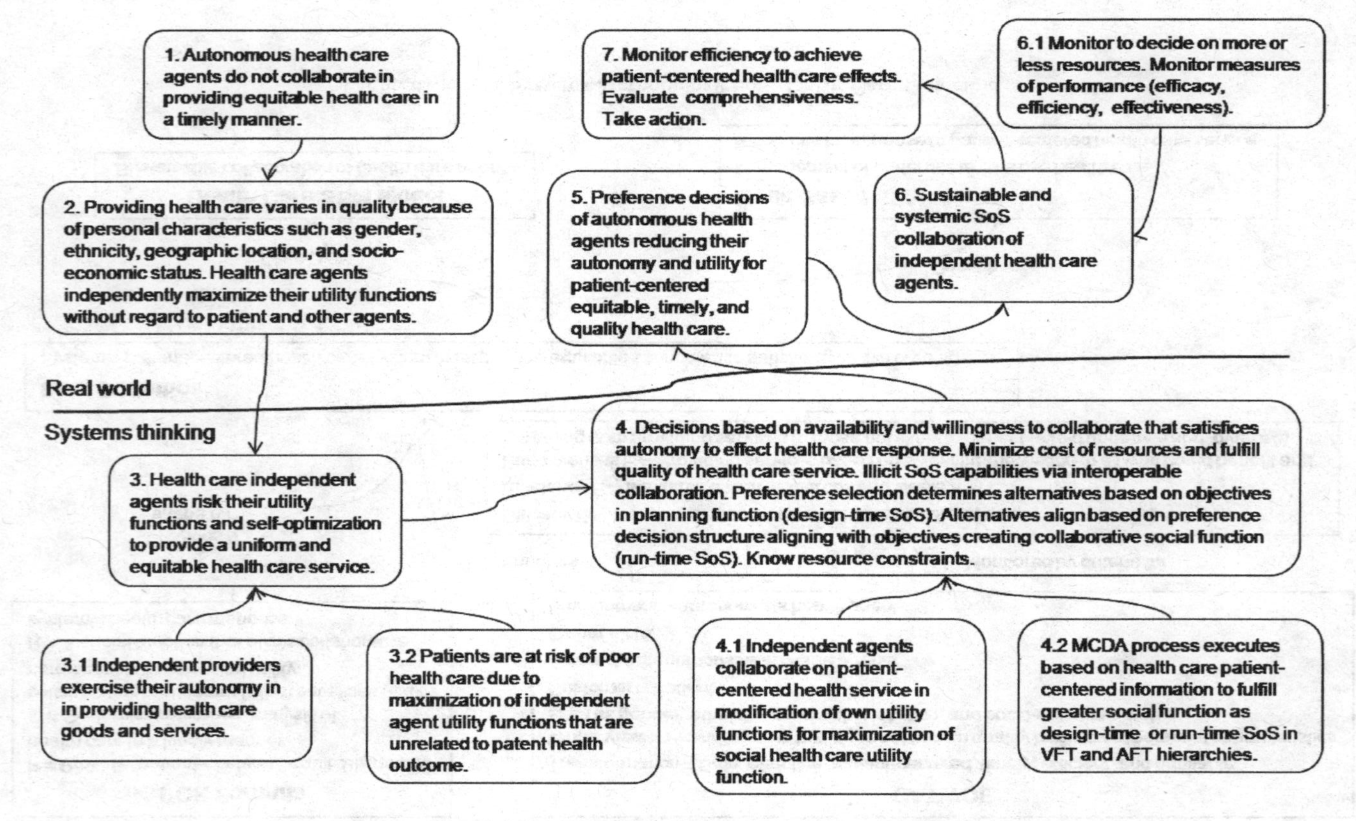

Fig. 32. Soft Systems Methodology Health Care SoS Model.

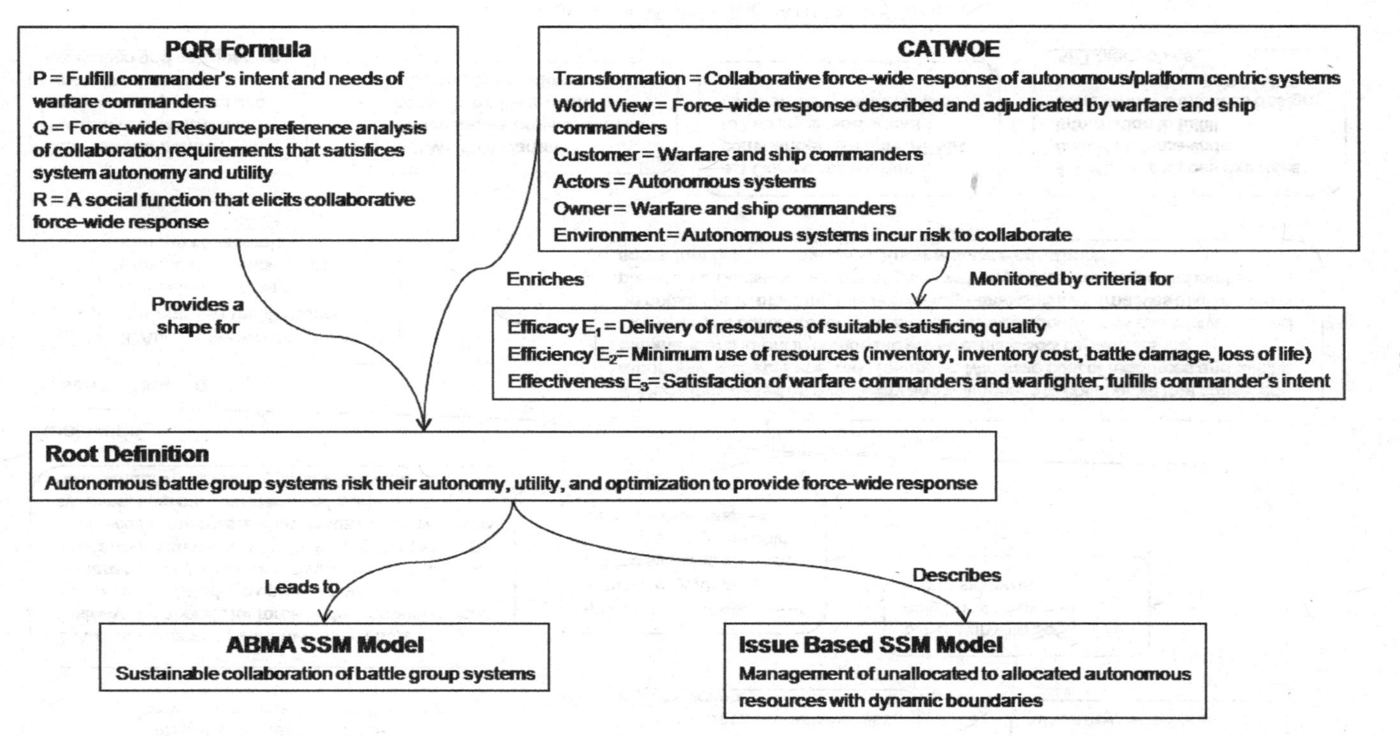

Fig. 33. Soft Systems Methodology SoS ABMA Guideline.

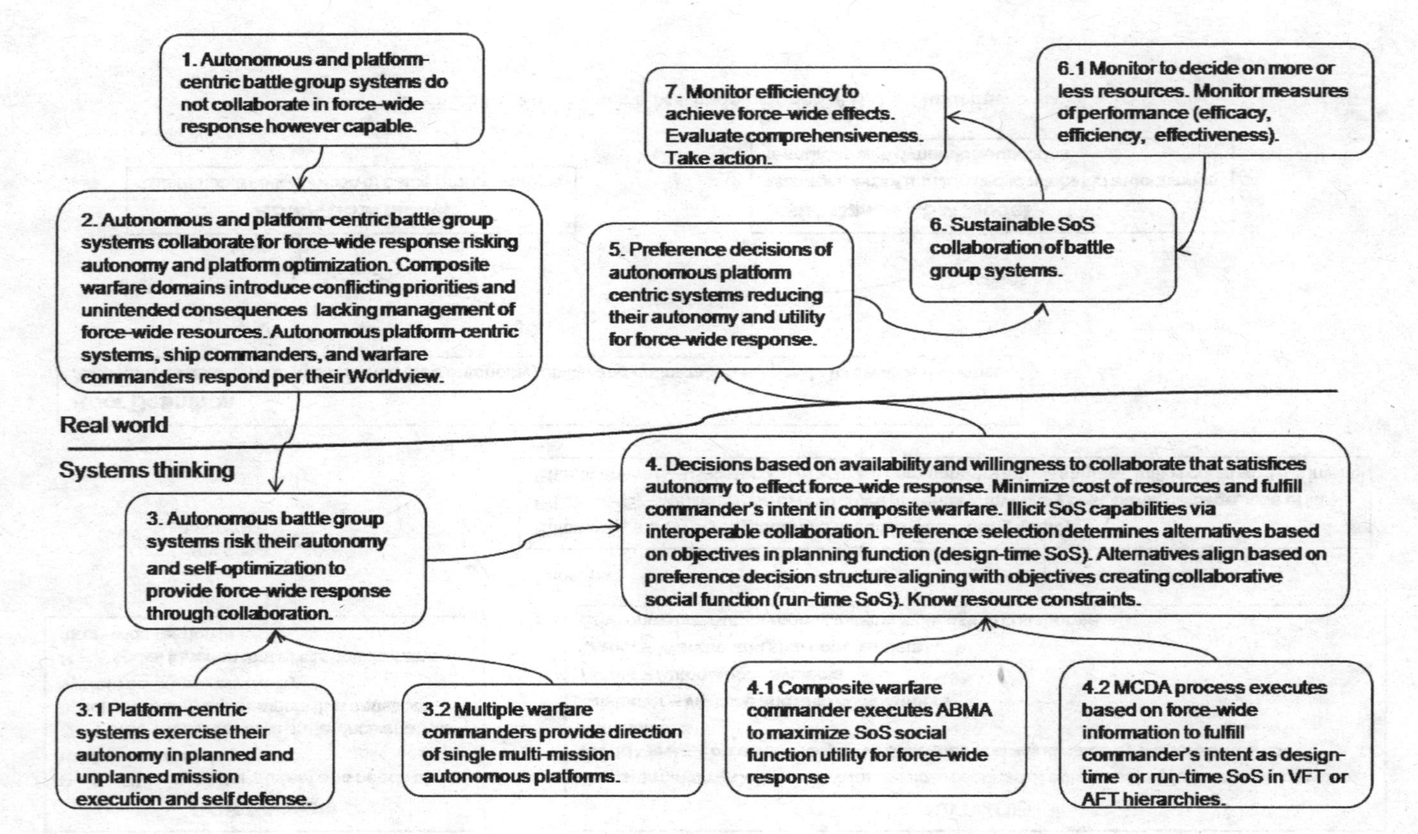

Fig. 34. Soft Systems Methodology ABMA SoS Model.

Appendix C

A CASE FOR AN INTERNATIONAL CONSORTIUM ON SYSTEM-OF-SYSTEMS ENGINEERING — IEEE SYSTEMS JOURNAL, SEPTEMBER 2007

D. DELAURENTIS
Purdue University, West Lafayette, IN, USA — corresponding author
ddelaure@purdue.edu

C. DICKERSON
BAE Systems, USA

M. DIMARIO
Lockheed Martin Corporation, Moorestown, NJ, USA

P. GARTZ
Boeing Company, Seattle, WA, USA

M. JAMSHIDI
The University of Texas, San Antonio, TX, USA

S. NAHAVANDI
Deakin University, Geelong, VIC, Australia

A. SAGE
George Mason University, Fairfax, VA, USA

E. SLOANE
Villanova University, Philadelphia, PA, USA

D. WALKER
Aerospace Corporation, Washington, DC, USA

A System of systems (SoS) conceptualization is essential in resolving issues involving heterogeneous independently operable systems to achieve a unique purpose. Successful operation as an SoS requires communication among appropriate individuals and groups across enterprises through an effective protocol. This paper presents a position on the creation of a consortium of concerned system engineers and scientists worldwide to examine the problems and solutions strategies associated with SoS. The consortium could lead efforts in clarifying ambiguities and in seeking remedies to numerous open questions with respect to SoS analysis, SoS Engineering (SoSE), as well as differences between Systems Engineering (SE) and SoSE. The mission of this consortium is envisioned to (i) act as a neutral party; (ii) provide a forum to put forth Calls to Action, and (iii) establish a community of interest to recommend set of solutions.

Index Terms: Academia, consortium, government, System of systems, IEEE, INCOSE, industry, military, system-of-systems (SoS).

Introduction

Recently, there has been a notably growing interest in System of Systems (SoS) concepts and strategies. The performance optimization among groups of heterogeneous systems in order to realize a common goal has become the focus of various application areas including military, security, aerospace, and disaster management [1–4]. There is particular interest in achieving synergy between these independent systems to enable the desired overall system performance [5]. In the literature, researchers have begun to address the issue of coordination and interoperability in a SoS [2, 5, 6] pointing to the emergence of the concept of system-of-systems engineering (SoSE). SoSE presents new challenges that are related to, but distinct from, systems engineering (SE) challenges. By understanding these differences, appropriate methods, tools, and standards can be crafted in an intelligent manner.

There has been a simultaneously recognition that significant organizational changes are needed in governments and industries, especially in the aerospace and defense areas, to realize these capabilities. The importance of having a group of systems working together as opposed to a single system is to increase the capability, robustness, and efficiency of data aggregation. In the US, major aerospace and defense manufacturers, including (but not limited to) Boeing, Lockheed-Martin, Northrop-Grumman, Raytheon, and BAE Systems, etc., all include some version of “large-scale systems integration” as a key part of their business strategies.

In some cases, these companies have even established entire business units dedicated to systems integration activities [2].

Proposed Consortium

Mission

The mission for the International Consortium for System of Systems (ICSOS) is to create a community of interest among science and engineering researchers and to foster proposals and solutions to advance the enhancement of SE to SoSE.

Objective

Ultimately, the objective of ICSOS is to enhance the ability of SoS community members to address many new and challenging problems facing the community across a diverse set of problem domain applications. The ICSOS will attempt to bring the best and brightest minds together each year in its annual workshop, creating a venue for funding agencies can obtain neutral views of the experts for their future program planning.

Application Areas of SoS

A misleading perception in the early days of SoS was that the application domain is exclusively information technology (IT) and, in particular, the "internet". This perception is of course far from the truth and one should simply look at the defense and aeronautics community in the US to note the many defense and security scenarios that fall naturally into SoS frameworks (e.g., future combat systems, unmanned air, sea, or land system of vehicles, etc.), as outline in the following.

National Security: SoS/SoSE began and is still primarily being applied within the US Department of Defense. Principle areas in which SoS/SoSE is finding early applicability are the national missile defense system of systems; Army's future combat systems development; and the interoperability and integration of Intelligence, Surveillance and Reconnaissance (ISR) systems and Command, Control, Computers, Communications and Information (C4I) systems and operations. However, it is beginning to be applied in non-defense or security related domains; such as, healthcare, transportation, and space exploration. Yet, any nations' security is dependent upon understanding all the environments of human endeavors and how they impact the people for whom they are responsible. Environments and systems that operate in and on them are growing in scale and complexity as humanity grows and matures. The concept of

SoS and the related SoSE endeavors are initiatives to develop processes, methods, and tools for understanding and dealing with the complexities and interactions of multiple domains and levels of uncertainty for addressing large-scale interdisciplinary problems and issues. Developing and sharing the SoS understandings and SoSE processes, methods, and tools within an international consortium forum will benefit both mankind and enhance a participating nations' security posture.

Even further, it is now quite obvious that the SoS concept has applicability beyond defense and IT applications. Here are some nondefense applications areas that ICSOS would also wish to bring to its annual workshop agenda.

Nondefense Applications of SoS

Civil Transportation: The National Transportation System (NTS) can be viewed as a collection of layered networks composed by heterogeneous systems for which the Air Transportation System (ATS) and its National Airspace System (NAS) is one part. At present, research on each sector of the NTS is generally conducted independently, with infrequent and/or incomplete consideration of scope dimensions (e.g. multi-modal impacts and policy, societal, and business enterprise influences) and network interactions (e.g. layered dynamics within a scope category). This isolated treatment does not capture the higher level interactions seen at the NTS or ATS architecture level; thus, modifying the transportation system based on limited observations and analyses may not necessarily have the intended effect or impact. A systematic method for modeling these interactions with a SoS approach is essential to the formation of a more complete model and understanding of the ATS, which would ultimately lead to better outcomes from high-consequence decisions in technological, socioeconomic, operational, and political policy making context [7]. This is especially vital as decision makers in both the public and private sector, for example, at the inter-agency Joint Planning and Development Office (JPDO) which is charged with transformation of air transportation, are facing problems of increasing complexity and uncertainty in attempting to encourage the evolution of superior transportation architectures [8].

Critical Infrastructure and Air Transportation Security: Air transportation networks consist of concourses, runways, parking, airlines, cargo terminal operators, fuel depots, retail, cleaning, catering, and many interacting people including travellers, service providers, and visitors.

The facilities are distributed and fall under multiple legal jurisdictions in regard to occupational health and safety, customs, quarantine, and security. Currently, decision making in this domain space is focused on individual systems. The challenge of delivering improved nation wide air transportation security, while maintaining performance and continuing growth, demands a new approach. In addition, information flow and data management are a critical issue, where trust plays a key role in defining interactions of organizations. SoS methodologies are required to rapidly model, analyse, and optimize air transportation systems [9]. In any critical real world system, there is and must be a compromise between increased risk and increased flexibility and productivity. By approaching such problem spaces from a SoS perspective, we are in the best position to find the right balance.

***SoS Standards*:** SoS literature, definitions, and perspectives are marked with great variability in the engineering community. It is also viewed as an extension of SE to a means of describing and managing social networks and organizations, the variations of perspectives leads to difficulty in advancing and understanding the discipline. Standards have been used to facilitate a common understanding and approach to align disparities of perspectives to drive a uniform agreement to definitions and approaches. By having the ICSOS represent to the IEEE and INCOSE for support of technical committees to derive standards for SoS will help unify and advance the discipline for engineering, healthcare, banking, space exploration, and all other disciplines that require interoperability among disparate systems.

***SoS and Healthcare Systems*:** Under a 2004 Presidential Order, the US Secretary of Health has initiated the development of a National Healthcare Information Network (NHIN), with the goal of creating a nationwide information system that can build and maintain Electronic Health Records (EHRs) for all citizens by 2014. The NHIN system, architecture currently under development, will provide a near real time heterogeneous integration of disaggregated hospital, departmental and physician patient care data, and will assemble and present a complete current EHR to any physician or hospital a patient consults [10]. The NHIN will rely on a network of independent Regional Healthcare Information Organizations (RHIOs) that are being developed and deployed to transform and communicate data from the hundreds of thousands of legacy medical information systems presently used in hospital departments, physician offices, and telemedicine sites into NHIN-specified metaformats that can be securely relayed and reliably interpreted anywhere in the country. The NHIN "network of networks" will clearly be a very complex SoS, and the performance of the NHIN and RHIOs will directly affect the safety, efficacy, and efficiency of healthcare in the US. Simulation, modeling, and other appropriate SoSE tools are under development to help ensure reliable, cost-effective planning, configuration, deployment, and management of the heterogeneous, life-critical NHIN and RHIO systems and subsystems [11]. ICSOS represents an invaluable opportunity to access and leverage SoSE expertise already under development in other industry and academic sectors. ICSOS also represents an opportunity to discuss the positive and negative emergent behaviors that can significantly affect personal and public health status and the costs of healthcare in the US.

***Global Earth Observation SoS*:** GEOSS is a global project consisting of over 60 nations whose purpose is to address the need for timely, quality, longterm, global information as a basis for sound decision making [12]. Its objectives are: (i) improved coordination of strategies and systems for Earth observations to achieve a comprehensive, coordinated, and sustained Earth observation system or systems; (ii) a coordinated effort to involve and assist developing countries in improving and sustaining their contributions to observing systems, their effective utilization of observations, and the related technologies; and (iii) the exchange of observations recorded from *in situ*, air full, and open manner with minimum time delay and cost. In GEOSS, the "SoSE process provides a complete, detailed, and systematic development approach for engineering SoS. Boeing's new architecture-centric, model-based systems engineering process emphasizes concurrent development of the system architecture model and system specifications. The process is applicable to all phases of a system's lifecycle. The SoSE process is a unified approach for system architecture development that integrates the views of each of a program's participating engineering disciplines into a single system architecture model supporting civil and military domain applications" [12]. ICSoS will be another platform for all concerned around the globe to bring the progress and principles of GEOSS to formal discussions and examination on an annual basis.

***Engineering of SoS*:** As we have noted, there is much interest in the engineering of systems that are comprised of other component systems and where each of the component systems serves organizational and human purposes. These systems have several principal characteristics that make the system family designation appropriate: operational independence of the individual systems, managerial independence of the systems; often large geographic and temporal distribution of the individual systems; emergent behavior, in which the system family performs functions and carries out purposes that do not reside uniquely in any of the constituent systems but which evolve over time in an adaptive manner and where these behaviors arise as a consequence of the formation of the entire system family and are not the behavior of any constituent system. The principal purposes supporting engineering of these individual systems and the composite system family are fulfilled by these emergent behaviors. Thus, a SoS is never fully formed or complete. Development of these systems is evolutionary and adaptive over time, and structures, functions, and purposes are added, removed, and modified as experience of the community with the individual systems and the composite system grows and evolves. The systems engineering and management of these systems families poses special challenges. This is especially the case with respect to the federated systems management principles that must be utilized to deal successfully with the multiple contractors and interests involved in these efforts. Please refer to the paper by Sage and Biemer in this issue [14].

ICSOS Participants

Analogous to a SoS itself, the consortium is a construct for the coming together of independent organizations to achieve purposes greater than those possibly met by isolated action. The participant base of the proposed international consortium on SoS (ICSOS), is depicted in Fig. 1. The central role of the ICSOS is intended to imply that it will act as a neutral (and nonprofit) broker in translating the needs of Government(s) to the field of SoS researchers and foster teaming arrangement from industry and academia to respond to the challenges posed.Research and development issues such as standards, network centricity, simulation, modeling, system engineering tools for SoS, management of SoS, emergence, evolution, etc., are among issues which ICSOS can address in its annual international focused workshop. Among application areas of SoS, where ICSOS would address are national defense, homeland security, civil transportation, critical infrastructure, healthcare, energy and power grids, manufacturing, space exploration, etc. Other issues of concern for ICSOS could include technology transfer, commercialization of intellectual property in SOS, etc. Efforts will be made to enable active participation and collaboration among

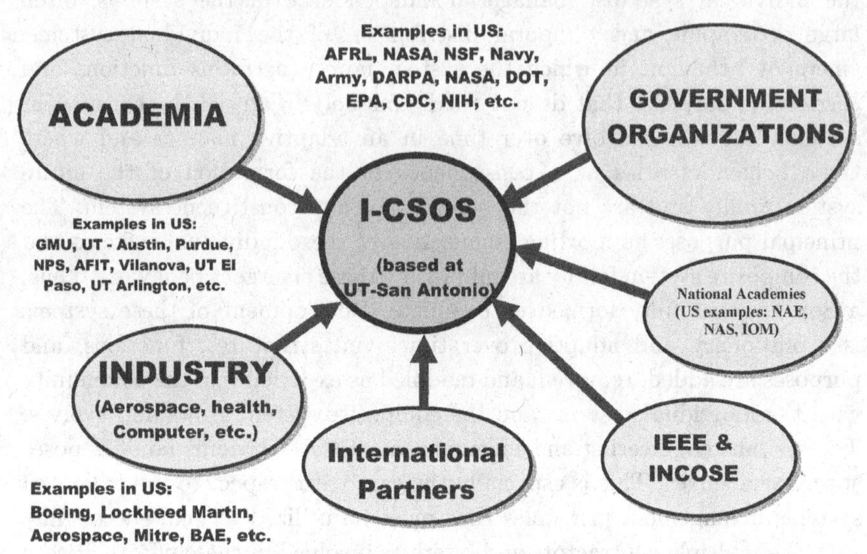

Fig. 1. Structure of International Consortium on System of systems — ICSOS.

IEEE and INCOSE members in this annual event and all ICSOS activities. A collaborative relation between these two organizations will establish a technically and geographically broad foundation for the promotion of SoS practices in SE. IEEE Systems Council and IEEE Systems, Man and Cybernetics have already approved technical co-sponsorship of ICSOS. This would make the annual workshop an IEEE technical sponsored event. An opportune time for the Workshop can be either a date concurrent with the annual IEEE SoSE Conference or an INCOSE event.

Conduct of ICSOS

Annual Workshop: ICSOS will have an annual workshop in each spring, whose technical agenda is directly on the demand of the members of the Consortium. Members will propose items for the annual workshop agenda and a steering committee would review it and approve or disapprove it for the official workshop agenda.

Finances: ICSOS is a not-for profit endeavor and will not have any resources except whatever its members are willing to help in keeping it a viable neutral party in meeting the challenges of SoS. We propose an optional membership fee structure for various categories of its members — from large corporations to medium size to small companies and among academia in tier I and tier II group of institutions.

Conclusions

This paper constitutes a white paper to propose the formation of a neutral scientific body to facilitate teaming of industrial and academic scientists and engineers to focus their efforts towards addressing many challenging problems facing the SoSE. This body is being called ICSOS. ICSOS will be a nonprofit body which will bring these groups of scientists and engineers on an annual workshop, which will take place after the annual IEEE SoSE Conference. The long-term effect of ICSOS is to draw the attention of governmental agencies towards addressing key issues and solve the associated challenging problems of SoS and SoSE. IEEE Systems, Man and Cybernetics (SMC) and the IEEE Systems Council (SysC) have both agreed to make the ICSOS annual workshop. The first annual ICSOS workshop will be held in Monterey, CA, on June 5, 2008.

Acknowledgements

The authors would like to thank many reviewers for their valuable comments, including Dr. Mark Johnson of the Aerospace Corporation, Washington, DC.

References

1. F. Sahin, P. Sridhar, B. Horan, V. Raghavan and M. Jamshidi, "System of Systems Approach to Threat Detection and Integration of Heterogeneous Independently Operable Systems," accepted, *IEEE Conference on Systems, Man and Cybernetics*, Montreal, Canada, October 2007.
2. Crossley, W. A., "System of Systems: An Introduction of Purdue University Schools of Engineering's Signature Area," *Engineering Systems Symposium*, Tang Center, Wong Auditorium, MIT, March 29-31 2004,
3. Lopez, D., "Lessons Learned From the Front Lines of the Aerospace," *Proc. of IEEE International Conference on System of Systems Engineering*, Los Angeles, April 2006.
4. Wojcik, L., A., Hoffman, K. C., "Systems of Systems Engineering in the Enterprise Context: A Unifying Framework for Dynamics," *Proc. of IEEE International Conference on System of Systems Engineering*, Los Angeles, April 2006.
5. Abel, A. and Sukkarieh, S., "The Coordination of Multiple Autonomous Systems using Information Theoretic Political Science Voting Models," *Proc.*

of *IEEE International Conference on System of Systems Engineering*, Los Angeles, April 2006.
6. DiMario, M., J., "System of Systems Interoperability Types and Characteristics in Joint Command and Control," *Proc. of IEEE International Conference on System of Systems Engineering*, Los Angeles, April 2006.
7. DeLaurentis, D.A., "Understanding Transportation as a System-of-Systems Design, Problem," *AIAA Aerospace Sciences Meeting and Exhibit*, 10–13 Jan. 2005. AIAA-2005-123.
8. DeLaurentis, D.A., Callaway, R.K., "A System-of-Systems Perspective for Future Public Policy," *Review of Policy Research*, Vol. 21, Issue 6, Nov. 2006.
9. Nahavandi, S., "Modeling of large complex system from System of Systems perspective", Keynote presentation, *2007IEEE SoSE Conference*, San Antonio, USA, 18 April 2007.
10. E. Sloane. "Understanding the Emerging National Healthcare IT Infrastructure." 24 × 7 *Magazine*. December, 2006.
11. E. Sloane, T. Way, V. Gehlot and R. Beck. "Conceptual SoS Model and Simulation Systems for A Next Generation National Healthcare Information Network (NHIN-2)", *Proceedings of the 1st Annual IEEE Systems Conference*, Honolulu, HI, April 9-12, 2007.
12. J. Pearlamn, "GEOSS — Global Earth Observation System of Systems," Keynote presentation, *2006 IEEE SoSE Conference*, Los Angeles, CA, USA, April 24, 2006.
13. M. L. Butterfield, J. Pearlman, and S. C. Vickroy, "System-of-Systems Engineering in a Global Environment," *Proceedings of International Conference on Trends in Product Life Cycle, Modeling, Simulation and Synthesis PLMSS*, 2006.
14. A. P. Sage and S. M. Biemer, "Processes for System Family Architecting, Design, and Integration," *IEEE Systems Journal*, ISJ1-1, September, 2007.

Dan DeLaurentis is an Assistant Professor with the School of Aeronautics and Astronautics at Purdue University, West Lafayette, IN, since 2004 under Purdue's College of Engineering Signature Area in System-of-Systems. His research interests include specific applications of interest in the aerospace sector. His recent research has been focused upon the development of an intellectual foundation for addressing problems characterized as system-of-systems. Pursuit of this focus includes establishment of an effective frame of reference, crafting of a common lexicon, and understanding of distinguishing behaviors and dynamics. Since he approaches these problems from a *design perspective*, the foundation is used to develop methods and tools for modeling and analysis which include especially probabilistic robust design (including uncertainty modeling/management), agent-based modeling, object oriented simulations, network topology analysis, and numerical and visual tools for capturing the interaction of system requirements, concepts, and technologies. The context for the research has been the exploration of future aerospace architectures, especially including the presence of innovative vehicles, new business models, and alternate policy constructs. In the past two years, his research has been sponsored under grants from the NASA Langley Research Center (concerning the development of future National Transportation Architectures via a System-of-Systems approach). He was also co-organizer for a Workshop held in May, 2006 entitled "*Methods for Designing, Planning, and Operating System-of-Systems*," sponsored by the Air Force Office of Scientific Research (AFOSR).

Charles Dickerson received the PhD degree from Purdue University, Lafayette, IN, in 1980.

He is a Technical Fellow with BAE Systems, Reston VA, who provides corporate leadership for architecture-based and system of systems engineering (SoSE). He has authored numerous papers and co-authored key reports. Currently, as Chair of the Architecture Working Group, International Council on Systems Engineering, he is working with the international community on architecture and standards, to include supporting the Office of the Secretary of Defense on SoSE guidelines. Before joining BAE, as a staff member at Massachusetts Institute of Technology (MIT), Lincoln Laboratory, Cambridge, he served in an IPA position as a Director of Architecture for the Chief Engineer of the Assistant Secretary of the Navy for Research, Development, and Acquisition. As an MIT IPA, he previously had served as Aegis Systems Engineer for the Navy Theater Wide Program. He also conducted research and directed flight tests at Lincoln Laboratory for low-altitude radar propagation and electronic countermeasures. His aerospace experience includes advanced air vehicle design and survivability at Lockheed's Skunk Works and Northrop's Advanced Systems Division. He has also served on the professional staff of the Center for Naval Analyses.

Michael DiMario (M'88) received the B.S. degree in natural science and the M.S. degree in computer science from the University of Wisconsin, Madison, and the MBA degree in management of technology from the Illinois Institute of Technology, Chicago. He is currently pursuing the PhD degree in systems engineering from Stevens Institute of Technology, Hoboken, NJ.

He is a Senior Program Manager with Lockheed Martin Corporation in Moorestown, NJ where he manages C4ISR and advanced technology programs. Prior to joining the program management organization, he served as Director of Systems and Software Quality. Prior to joining Lockheed Martin in 2001, he spent 19 years with AT&T subsequently becoming Lucent Bell Laboratories, in Naperville, IL, and Murray Hill, NJ, in managerial and director leadership roles for software development and systems engineering of switching and broadband telecommunication systems, software test and verification, and total quality management. During his tenure at Lucent Bell Labs, Michael led numerous ISO 9001 and SEI certifications, and Malcolm Baldrige applications, as well as the creation of the Total Quality Management, and systems engineering business models and corporate infrastructures. He was instrumental in the creation of five telecommunications software engineering Lucent Best Current Practices, TL9000 telecommunications standard, managed the scaling of systems engineering and software development processes for new acquisitions, and led major telecommunication outage causal teams. He is the author of numerous papers and articles on software quality and systems engineering. Mr. DiMario is an active member in the INCOSE.

Paul Gartz (SM'89) is a System-of-Systems Architect and member of the Boeing Technical Fellowship, Boeing, Seattle, WA. He was Chief Archi tect for NOAA SoS, which evolved into a GEOSS focus, and head airplane architect for a new Boeing wide-band SoS business called CbB. He led Boeing's first commercial systems engineering groups and was part of the development teams for most modern Boeing commercial transports. He has presidency positions in two IEEE business units namely "IEEE Operating Units."

Mr. Gartz is Past President of the Aerospace and Electronic Systems Society (AESS) and President-elect of the IEEE Systems Council. He reorganized AESS to expand globally and into systems-of-systems and created new officers for each major population and economic center in the world and began events of interest to in ATM, SoSs, Defense, UAVs, earth observation, sensors, and more.

Mo M. Jamshidi (Fellow-IEEE, ASME, AAAS, NYAS, TWAS, SM-AIAA) received his PhD Degree in Electrical Engineering from the University of Illinois at Urbana-Champaign in February 1971. He holds Honorary Doctorate Degrees from Azerbaijan National University, Baku, Azerbaijan (1999), the University of Waterloo, Canada (2004)and Technical University of Crete, Greece, (2004). Currently, he is the Lutcher Brown Endowed Chaired Professor at the University of Texas at the San Antonio Campus, San Antonio, Texas, USA. He was the founding Director of the Center for Autonomous Control Engineering (ACE) at the University of New Mexico (UNM) and moved to the University of Texas, San Antonio in early 2006. He has been the Director of the ICSoS — International Consortium on System of Systems Engineering (*icsos.org*) — since 2006. Mo is an Adjunct Professor of Engineering at Deakin University, Australia. He is also Regents Professor Emeritus of ECE at UNM. In 1999, he was a NATO Distinguished Professor in Portugal of Intelligent Systems and Control. Mo has written over 550 technical publications including 60 books and edited volumes. He is the Founding Editor, Co-founding Editor or Editor-in-Chief of 6 journals. Mo is Editor-in-Chief of the new *IEEE Systems Journal* and founding Editor-in-Chief of the *IEEE Control Systems Magazine*. Mo is a Member of Senior Member of the Russian Academy of Nonlinear Sciences, Hungarian Academy of Engineering and AIAA. He is a recipient of the IEEE Centennial Medal and IEEE CSS Distinguished Member Award. Mo is currently on the Board of Governors of the IEEE Society on Systems, Man and Cybernetics and the IEEE Systems Council. In 2005, he was awarded the IEEE SMC Society's Norbert Weiner Research Achievement Award and in 2006 the IEEE SMC Distinguished Contribution Award. In 2006 he was awarded a "Distinguished Alumni in Engineering" at the Oregon State University, Corvallis, OR.

Saeid Nahavandi received BSc (Hons), MSc and a PhD in Automation and control from Durham University (UK). Professor Nahavandi holds the title of Alfred Deakin Professor, Chair of Engineering and is the leader for the Intelligent Systems research Centre at Deakin University (Australia). He won the title of Young Engineer of the Year for in 1996 and has published over 300 peer reviewed papers in various International Journals and Conference and is the recipient of six international awards in Engineering. He Designed the World's first 3D interactive surface/motion controller. His research interests include modeling of complex systems, simulation based optimization, robotics, hap tics and augmented reality. Professor Nahavandi is the Associate Editor — IEEE Systems Journal, Editorial Consultant Board member International Journal of Advanced Robotic Systems, Editor (South Pacific region) — International Journal of Intelligent Automation and Soft Computing. He is a Fellow of Engineers Australia (FIEAust) and IET (FIET) and Senior Member of IEEE (SMIEEE)

Andrew P. Sage — received the BSEE degree from the Citadel, the SMEE degree from MIT and the PhD from Purdue, the latter in 1960. He received honorary Doctor of Engineering degrees from the University of Waterloo in 1987 and from Dalhousie University in 1997. He has been a faculty member at several universities including holding a named professorship and being the first chair of the Systems Engineering Department at the University of Virginia. In 1984 he became First American Bank Professor of Information Technology and Engineering at George Mason University and the first Dean of the School of Information Technology and Engineering. In May 1996, he was elected as Founding Dean Emeritus of the School and also was appointed a University Professor. He is an elected Fellow of the Institute of Electrical and Electronics Engineers, the American Association for the Advancement of Science, and the International Council on Systems Engineering. He is editor of the John Wiley textbook series on Systems Engineering and Management, the INCOSE Wiley journal Systems Engineering and is coeditor of Information, Knowledge, and Systems Management. He edited the IEEE Transactions on Systems, Man, and Cybernetics from January 1972 through December 1998, and also served a two year period as President of the IEEE SMC Society. In 2000, he received the Simon Ramo Medal from the IEEE in recognition of his contributions to systems engineering and an IEEE Third Millennium Medal. In 2002, he received an Eta Kappa Nu Eminent Membership Award and the INCOSE Pioneer Award. He was elected to the National Academy of Engineering in 2004 for contributions to the theory and practice of systems engineering and systems management. His interests include systems engineering and management efforts in a variety of application areas including systems integration and architecting, reengineering, engineering economic systems, and sustainable development.

Elliot B. Sloane (Senior Member, IEEE) is on the faculty of Villanova University's School of Business, and has a BS-BME from Cornell University, and both an MSEE and a PhD in Information Science and Technology from Drexel University. Since 2000, he has been a prolific clinical engineer and information technology researcher, author, and lecturer who brings over three decades of active healthcare technology, manufacturing and information industry, and academic experience. He is serving his second three-year term on the IEEE EMBS Society Board of Directors, served as the chair of IEEE's Membership Development Committee on Healthcare Information Technologies from 2004–2005, and is the IEEE SA 11073 Medical Device Standards Sponsor. He is a past president of the American College of Clinical Engineering (ACCE), has co-chaired the joint ACCE/HIMSS Integrating the Healthcare Enterprise (IHE) Patient Care Device Domain program since 2003, and was recently appointed to the HIMSS Privacy and Security Steering Committee. In 2006 he was appointed co-chair of the overall IHE International Strategic Development Committee. Since 2006, he has also served on the ANSI Healthcare Information Technology Standards Panel, which is under contract with the US Secretary of Health to identify and recommend the best standards and practices for implementing the US National Healthcare Information Network (NHIN). On behalf of WHO and PAHO, he has taught in developed and developing countries around the world. He was awarded the ACCE Clinical Engineering Advocacy Award in 2006 and the joint ACCE/HIMSS CE-IT Synergies Award in 2007.

Donald R. Walker is senior vice president of Systems Planning and Engineering. He assumed this position on January 1, 2004. He is responsible for promoting the efficient use of corporate and government resources through horizontal and cross-program integrated planning and engineering. Systems planning and Engineering comprises. The Aerospace Corporation's Missile Defense Division, Colorado Operations, Office of the Chief Architect/Engineer, National Space Systems Engineering, Integrated System Architecture Office, and Strategic Awareness and Policy organizations. Walker, who joined Aerospace in September 2002, was vice president of National Space Systems Engineering before he was appointed to his current position. As vice president of National Space Systems Engineering he provided strategic support to senior-level Department of Defense customers in the development of an integrated national-security space capability. He is located in the company's Rosslyn, Virginia, office. Before joining The Aerospace Corporation Walker was an independent consultant to the government on intelligence and defense space systems. In 29 years with the Air Force, Walker acquired top-level leadership experience in the acquisition, launch, and operation of defense and intelligence space systems for the Air Force and the National Reconnaissance Office. He retired from the Air Force at the rank of brigadier general in 1995 and joined United Services Automobile Association (USAA), a Fortune 200 financial services company, as chief information officer.

Appendix D

SYSTEM OF SYSTEMS INTEROPERABILITY TYPES AND CHARACTERISTICS IN JOINT COMMAND AND CONTROL, IEEE CONFERENCE ON SYSTEM OF SYSTEMS ENGINEERING, APRIL 2006

MICHAEL J. DIMARIO
Department of Systems Engineering and Engineering Management
Stevens Institute of Technology Hoboken, NJ USA
dimario@patmedia.net

The United States Department of Defense and their Joint Vision 2010 and 2020 articulates the necessity for multinational and interagency cooperation and interdependent command and control for a full range of capabilities. Joint Vision relies on system of systems to bridge participant systems with interoperability as a key enabler. To achieve this key enabler, this paper discusses the need for interoperability implementation at the programmatic, constructive, and operational levels. The system of systems interoperability must take into account its unbounded characteristics and engineer constituent systems for syntactic and semantic interoperability and to be system of systems enabled for an omnipresent environment. This paper demonstrates why systems must be managed and engineered as system of systems centric versus systems centric. System Readiness Levels and Integration Readiness Levels are suggested to mange the constituent systems within a system of systems.

Keywords: System of systems, interoperability, command and control, Joint Vision, syntactic, semantic, convergence, SoS enabled, Technology Readiness Level, System Readiness Level, Integration Readiness Level, unbounded, bounded.

Introduction

The United States Department of Defense (DoD) and their Joint Vision 2010 (JV2010) and 2020 (JV2020) describes the US military and its goals for the 21st century. This vision assumes and articulates the necessity for multinational and US interagency cooperation and interdependent joint command and control (JC2) for a full range of capabilities operationally, organizationally, doctrinally, technically, and intellectually [1, 2]. Information superiority (IS) is a key enabler of the Joint Vision (JV) and the evolution of JC2 as it integrates the core competencies of the multiservices into joint operations [3]. Joint Vision is the optimal integration of all joint forces which in turn creates additional complexity to operations. It is this complexity and interdependent capabilities that will achieve full spectrum dominance through independent multinational and multiagency operations acting in concert.

Acting in concert and remaining independent requires a system of systems (SoS) whereby the independent agencies and multinational systems dynamically merge to form a holistic system. The result is a SoS that has greater capabilities than the sum or simple combination of the individual multinational or agency systems. JV relies on SoS to bridge between existing legacy and new systems. In order to connect these systems, interoperability is a key enabler for SoS to be realized for the JV. Interoperability is defined by the JV2020 as "the ability of systems, units, or forces to provide services to and accept services from other systems, units or forces and to use the services so exchanged to enable them to operate effectively together" [2]. Information systems and equipment must access shared networks as a recognized and allowed participant. It is recognized by the JV that interoperability is essential but not sufficient. The result of interoperability must be able to produce capabilities of a SoS that is greater than the sum of their individual constituent's capabilities.

SoS and Types of Interoperability

It is reasonable to assume that interoperability to support a SoS must also support the characteristics of a SoS. Not doing so would only create a larger system and the constituent systems would not be independent systems, but subsystems. The literature describes several SoS characteristics. Characteristics defined by Maier are operational element independence, managerial independence, evolutionary development, emergent behavior, and geographic distribution [4]. In the context of JC2 and interoperability, the SoS characteristics of autonomy, belonging, connectivity, diversity, and emergence [5] are similar and relevant. Both sets of characteristics as well as the myriad of SoS definitions describe existing or future systems and have their origins due to different perspectives of the SoS problem. A recent SoS study and survey by the US Air Force (USAF) has revealed that 75% of those surveyed defined a SoS as a large system comprised of many subsystems while 20% of surveyed respondents believe a SoS is comprised of cooperating specially built systems. The remaining 5% of those surveyed had the perspective that a SoS is comprised of a set of systems that interact dynamically and are responsive [6]. It is this latter perspective that is required by the Air Force and may be required by the JV. Although the two sets of characteristics are similar, the critical characteristics and importance to interoperability is connectivity, belonging, and diversity. This is inferred from a systemigram of the USAF C2 Constellation which provides a profound overview of the SoS interoperability (SoSI) characteristics [7].

Interoperability models define three types of interoperability, shown in Figure 1, within a SoSI context — programmatic, constructive, and operational. Programmatic interoperability concerns the linkages between program offices to manage system acquisition. Constructive interoperability concerns linkages between organizations that are responsible for system construction that entail architecture, standards, design, commercial off the shelf products (COTS), and maintenance. Operational interoperability concerns linkages between systems in their operation or technical interoperability, their interactions with one another, the environment, and with users.

The SoSI types are a collection of models derived from other interoperability models. System operation was defined by the Command, Control, Communications, Computer, Intelligence, Surveillance, and Reconnaissance (C4ISR) Levels of Information System Interoperability

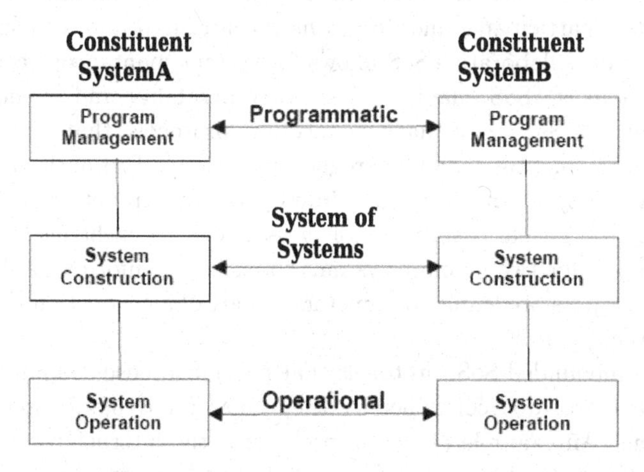

Fig. 1. System of Systems Interoperability Types [8].

(LISI) and further refined by adding Coalition Interoperability [9]. LISI uses a common frame of reference and measures of performance. It facilitates a common view of interoperability and the suite of capabilities enabling each logical level of system participant interaction. LISI addresses operational interoperability or semantic understanding along with technical interoperability or syntactic understanding between systems.

SoS Connectivity, Diversity and Interoperability

SoS connectivity is defined as "dynamically supplied by constituent systems with every possibility of myriad connections between constituent systems, possibly via a net-centric architecture, to enhance SoS capability" [5]. The connectivity, platform diversity, and associated interoperability defines whether the SoS is bounded or unbounded. In most SoSs and in the perceptions of the Air Force's surveyed respondents, bounded or directed systems are the majority of SoSs as they are characterized by centralized data and control. The linkages presuppose that systems are known a priori about each system participant and are developed to connect and interoperate. These systems also cost a great deal to add additional capabilities to interoperate between systems as connections and applications must be added to all affected systems. The challenge is to develop a SoS to accommodate capabilities or emergent behavior that is required in the future without high acquisition cost.

Characteristics of unbounded systems are that they must operate within a dynamic environment as they have an unknown number of collaborative participants and do not have centralized data or control. An unbounded or collaborative SoS allows for system spontaneous connection or the system is "SoS enabled." The interoperability and connection of an unbounded SoS must have connection protocols that simplify the connections among the many heterogeneous systems. Protocols are used to manage a variety of connection and interoperability properties. A protocol is defined as a formal description of messages to be exchanged and rules to be followed for two or more systems to exchange information [10]. The system must pass predefined parameters as part of a connection handshake between the systems.

In an unbounded SoS, protocols that provide a loose coupling and are omnipresent are protocols whereby they allow for dynamic spontaneous connections. An example of such a protocol is the internet protocol (IP). These protocols are denoted "convergent" since many users or systems have adopted a single rule to connect with other systems and the community has converged upon a universal protocol. Another example is the "protocol" and standards for domestic electricity supply of 120 volts at 60 HZ of alternating current with standard sockets. With this methodology, any electrical system may be connected to the electrical grid power system [6]. The collections of physical and electrical properties comprise the protocol or set of rules. These sets of rules may also be considered a standard and is how standards are eventually established. The use of protocols defines the expected rules of behavior among the systems of a SoS.

Similar to traditional distributed systems, the constituent systems in an unbounded domain can only connect with their neighbors. Survivability and reliability in such an environment is paramount and yet must remain flexible. Emergent algorithms defined as any computation that achieves formally or stochastically predictable global effects, by communicating directly with only a bounded number of immediate neighbors and without the use of central or global visibility [11] may be used in this dynamic arena to support interconnections that must adapt to changing conditions. Every node must be able to communicate with its neighbors, but do so without full knowledge of the topology of the network. Distributed algorithms distribute parts over the network and require resources in each node proportional to the size of the network.

An example of syntactic interconnection in an unbounded system is Mobile Ad Hoc Networks (MANETs). A MANET is a set of wireless mobile devices that self organize into a network without relying on central control or a fixed infrastructure. All nodes are equal and join or leave the network at will. The nodes also offer the capability to route data for one another in a multi-hop scenario [12]. An example of semantic interoperability is the Joint Single Integrated Air Picture (SIAP) Systems Engineering Organization (JSSEO) that is building an Integrated Architecture Behavior Model (IABM). The IABM is a Platform Independent Model (PIM) to provide SIAP functionality to DoD assets that provide tracking capabilities in accordance with the High Level Architecture (HLA) standard. The current problem to be solved by the IABM is that existing assets are able to share tactical data syntactically but differ in their interpretation of the data semantically. The IABM will resolve this problem by providing the same interpretation of SIAP data in a heterogeneous SoS [13].

For information systems or the JC2, the interoperability between systems requires systems to exchange information syntactically or have the ability to exchange information. However, to exchange the data does not imply proper usage of the information. The semantics of the information or the ability to understand and act upon the information must also be understood by the recipient system. Thus syntactic and semantic interoperability is critical for a SoS to properly connect and interoperate [14]. Such a SoS operational interoperability type would have the MANET syntactic characteristics and the IABM semantic characteristics.

SoS Belonging and Interoperability

An additional characteristic of an unbounded SoS is the constituent systems choose to belong to the greater collective and are dependent on the value of the union. The capabilities of the participant and the expected capabilities and emergent behavior of the interoperability union in SoS define the characteristic of belonging. This is in contrast to strictly a system whereby its elements do not have a choice, belonging is non-negotiable, and are captive fulfilling a specific objective for the system [15]. In a bounded system, belonging is directed and specifically designed to interoperate without choice.

For purpose of security and trustworthiness in unbounded and bounded SoSs, systems may not be allowed to interoperate even though they may have the capability to interoperate. Systems may be blocked from joining the SoS collective, as they may be perceived as being unfriendly, harbor the potential of unacceptable emergent behavior, or be restricted due to bandwidth considerations of exchanging data or for overload control. The value of the participant defines whether a system belongs, either permanently or occasionally, in a greater SoS collective [14].

JV SoS Command and Control

SoS Interoperability

The JV is supported by the Joint Interoperability Test Command (JITC) organization whose mission is to provide independent operational test and evaluation of C4I systems, identify C4I interoperability deficiencies, and joint C4I interoperability testing evaluation and certification. The Joint Distributed Engineering Plant (JDEP), a function of the JITC, is a DoD-wide capability for Service and Joint engineering, integration, and test resources to provide SoS battlefield representative environments in support of developer, tester, and warfighter requirements. JDEP interaction hierarchy Federation Object Model (FOM), shown in Figure 2, defines a

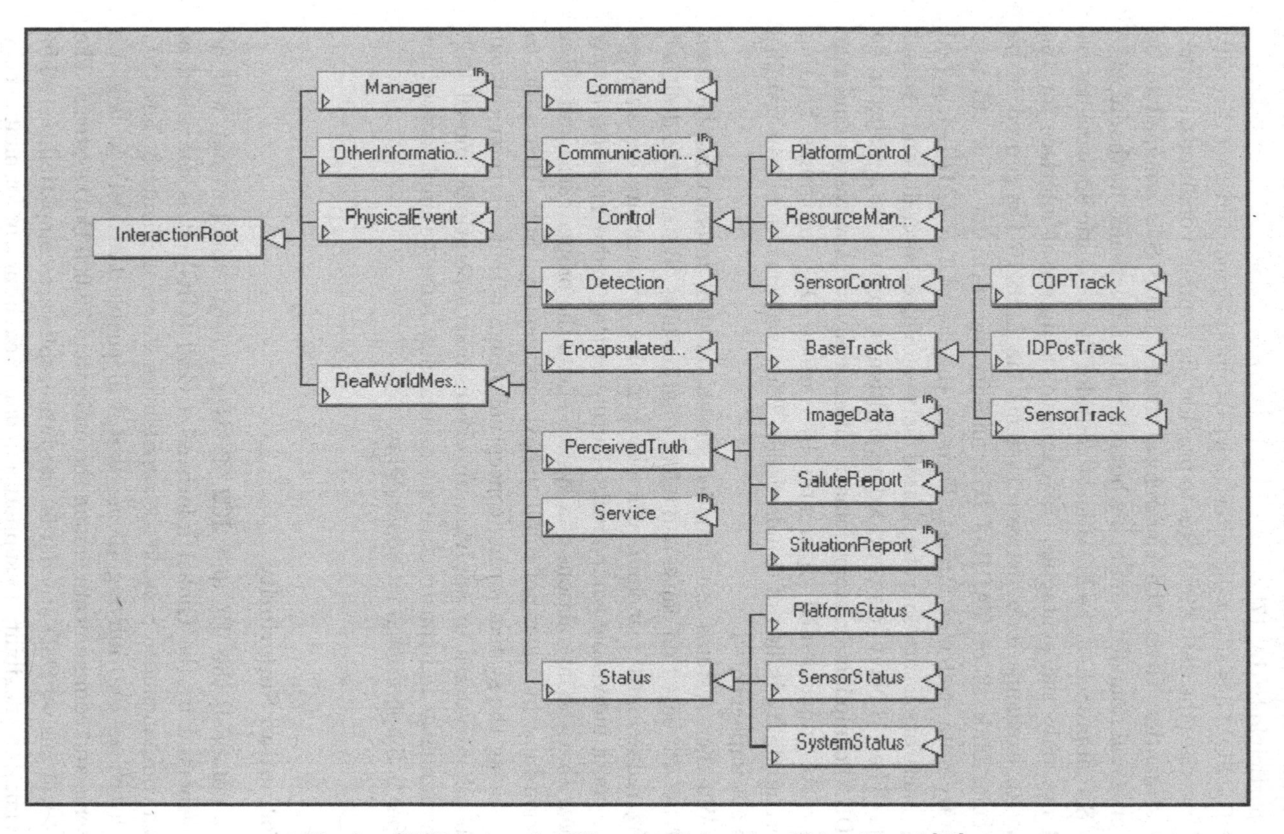

Fig. 2. JDEP Interaction Hierarchy Federation Object Model [17].

data model for exchange of persistent and transient information between SoS participants [16, 17].

The FOM has adaptable interfaces to support agency interface diversity and manages the data exchanges to provide the required conditions, syntax, and semantics. Given the hierarchical model, the SoS is assumed to be directed and bounded reinforcing bounded and reductionist methodologies. The SoS approach of the FOM must consider the semantic implementation and not only the syntactic. For example, a universal interface for the Navy to exchange data between systems via Link 16 has proved to be difficult. Link 16 or TADIL-J (US military synonym for Link 16) is a military data exchange format. Through the use of the TADIL-J, a real time tactical picture may be shared between ships, aircraft, and ground units. The US Navy has provided acquisition incentives for systems to be TADIL-J compliant or programmatic interoperability; however the message sets were implemented differently by various DoD programs limiting interoperability in lieu of standards [8]. Thus constructive interoperability was not implemented.

Whether systems are bounded or unbounded, interactions between systems are based on cause and effect relationships. Systems are designed using reductionist principles whereby complex systems are decomposed into sets of indivisible components. During testing and operations, it may appear that the interactions of a system elicits the proper cause and effect relationships, but does not fulfill the expectations of a SoS. This is due to not placing the system during conceptual design within a greater context with other systems or applying an expansionist operational model [18]. Programmatic and constructive interoperability must be implemented for proper SoS operational interoperability.

Information Superiority

To accomplish the JV and JC2, IS is a critical enabler that will be implemented by the Global Information Grid (GIG). It is comprised of seven components to provide the capabilities shown in Figure 3. The GIG is viewed as the unifying theme that will enable the DoD to field its systems and communicate among the more than 28,000 DoD systems. The GIG will ensure DoD standards, hardware, software compatibilities while providing the flexibility to support a dynamic environment for multiagency and multinational cooperation [3].

Warrior Component
- Networked planes, ships, vehicles, missiles, ground troops, platforms

Global Applications
- Mission/functional automated information systems

Computing
- Common user processing services

Communications
- Bandwidth on demand

Network Operations
- Network Management, Information Dissemination Management,, Information Assurance

Information Management
- Information on demand
- Life cycle management

Foundation
- Doctrine, Architecture, Standards, Policies, Organization, Resourcing, Training, Testing, Governance

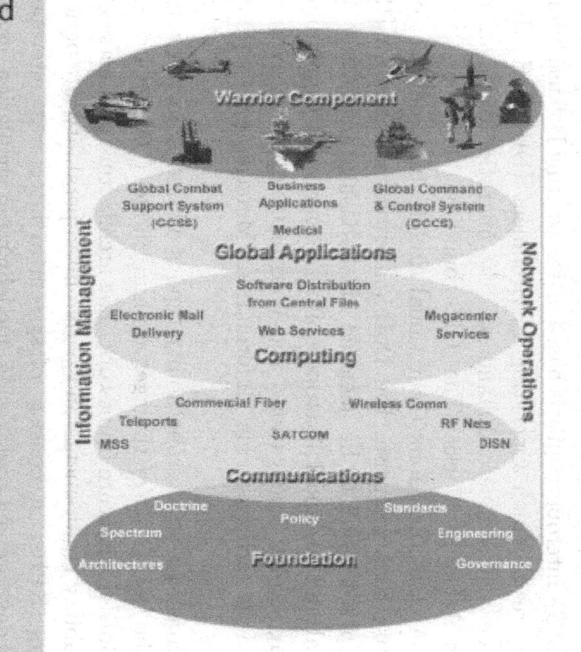

Fig. 3. Global Information Grid Components [3].

The GIG's interoperability is based on the existing Defense Information Infrastructure Common Operating Environment (DII COE). It is also based on current commercial standards, the migration of legacy systems, and other building blocks. The GIG building blocks are the Joint Technical Architecture, Joint Operational Architecture, and Joint Systems Architecture that comprises the DoD enterprise architecture [19]. The GIG vision is for SoS to be achieved via horizontal integration through operational and technical interoperability through the Communications, Computing, Global Applications, and Warrior Component levels. Vertical integration is achieved via Information Management and Network Operations.

The constructive interoperability type is addressed in the Foundation level via the DoD Office of Management and Budget (OMB) Circular A-130 and the DoD architecture framework. The architecture products document interoperability requirements and derive interoperability key performance parameters (KPPs). It consists of system architectures, interoperability standards development and testing, modeling and simulation, engineering, and data standardization. The programmatic interoperability type is not evident in the literature and is largely absent in practice [20]. However, the OMB Circular A-130 states that enterprise architecture must include a description of the business or operational processes [19]. The Foundation will have to include programmatic interoperability for successful SoS implementation in the GIG. DoD acquisition processes should be included in order to align the programmatic interoperability type to properly drive the constructive interoperability type. The development and maintenance life cycle is fluid between programmatic, constructive, and operational interoperability types in the GIG's systems. It is this fluid nature that stresses the importance of solid SoSI types and the implementation of their respective processes.

SoS implementation will prove to be difficult unless lessons learned from previous SoSI evaluations and JTIC evaluations are not embraced. For example, in 1998 the US Navy Cooperative Engagement (CEC) was fielded, at a program cost of $3.5 billion. The CEC is a major contributor to the JV 2010 concept of full-dimensional protection for the fleet from air threats as a major system with a fleet's SoS [21]. CEC is a system of hardware and software that allows for sharing of radar data on air targets among other CEC equipped ships. A ship within the Battle Group may launch an anti-air missile at an air threat or anti-ship cruise missile based on track data sent from another ship. Interoperability problems delayed CEC deployment as a result of technical evaluations, known as TECHEVALs which is equivalent to SoSI syntactic evaluation. An operational evaluation known as OPEVAL and equivalent to SoSI semantic testing in February of 2001 revealed improved CEC SoSI syntactically but required improved CEC SoSI semantics. The root cause determined by the US Navy of the problems encountered in developing and fielding CEC was the lack of programmatic and constructive interoperability and a ship-centric or system interoperability focus versus a SoSI or Battle Group centric focus [22].

JV SoS Evolution

The JV JC2 to be successful as a fully integrated SoS must solve the difficulties of the operational threads or paths between the participant systems. Problems of connectivity, capacity, utilization, overload, data latency, syntactic incompatibility, and undesirable semantic emergent behavior have been simply shown in the Link 16 experience. Trade offs between quality of service and the amount of information exchanged will have to be assessed. The participant system's interoperability must be regulated by security, functionality, value, and reliability. The importance of programmatic and constructive interoperability leading to poor operational interoperability is evident in the CEC example.

The primary SoSI challenges between legacy and new systems is to provide a maturity construct that allows for different levels of sophistication regarding the exchange of information and implementation options for the various levels. There also needs to be a comprehensive understanding of the interoperability attributes of information systems. Systems evolution onto itself manifests many problems and is only magnified in a SoS. The maintenance of the interoperability of evolving systems must be managed with the same attention as new systems are developed. The properties of the evolution and interoperability that affect how the systems in SoS evolve are provided and modified from [23]:

1. Relative stability of the components
2. Rate of change between systems
3. The number of interoperability SoS relationships
4. The number and complexity of interoperations affected by evolution
5. Coordination of system evolution with other systems
6. Commonality of evolution — systems must evolve in the same operational direction

7. Ability to assess trust in a changing system
8. Availability of components

The systems within the SoS will be of various maturity levels. The systems will be newly developed, COTS, legacy systems, and a myriad of hybrids. The technologies within the systems are likely to be assessed against Technology Readiness Levels (TRL). Systems Readiness Level (SRL) and Integration Readiness Level (IRL) indices that include the concept of the TRL scale with an established understanding of the relationship between them may be employed to manage JV technologies and systems as well as a complement to LISI. A SRL is comprised of five levels and is a function of the individual TRLs. The IRL is comprised of seven levels and is a systematic measurement of the interfacing of compatible interactions for various technologies and the consistent comparison of the maturity between integration points [24]. The IRL index takes into account the quality and complexity of the system interactions with other systems.

Conclusion

A SoS is not realizable if the systems cannot communicate. Communication is only the initial step of interoperability, as a system must know what to do with the information that is communicated adding value to the SoS. SoSI is fundamental to the SoS complex to provide value that is greater than the sum of its individual participants. The methodologies of SoS design and their associative interoperability concept of operations will define the SoSI characteristics. The characteristics will be different for bounded versus unbounded SoSs. High acquisition cost is expected for bounded systems. Unbounded systems are rare with little experience in acquisition costs. However, it is clear that for a true SoS, the participant systems interact dynamically and are responsive. They are "SoS enabled" whereby their interoperability is based on a convergence protocol. The convergence protocol is omnipresent and will be used when a system adds value or dynamically belongs to the SoS collective. The key point is for a group of system participants, through readily available interoperability schema to elicit value greater than their own value without syntactic change. Thus acquisition cost is low and reliability is high for SoSI.

The SoSI success for JV JC2 depends on how the DoD embraces and modifies its acquisition and systems engineering processes for programmatic, constructive, and operational interoperability. To date, SoSI has high acquisition cost and initial poor reliability. Improved SoSI systems engineering and interoperability syntactic and semantic engineering is required for systems to be "SoS enabled" and to transition from being systems centric to SoS centric. A significant challenge for SoSI is an increase in complexity without an increase in hierarchy, control, and acquisition cost.

References

1. Chairman of the Joint Chiefs of Staff, "Joint Vision 2010," June 1996. Washington, DC, U.S. Government Printing Office, www.dtic.mil/jv2010/jvpub. htm, Retrieved January 8, 2006.
2. Chairman of the Joint Chiefs of Staff, "Joint Vision 2020, June 2000," Washington, DC, U.S. Government Printing Office, www.dtic.mil/jointvision/jvpub2.htm, Retrieved January 8, 2006.
3. Chairman of the Joint Chiefs of Staff, "Enabling the Joint Vision," January 2000, Washington, DC, U.S. Government Printing Office, www.iwar.org.uk/rma/resources/jv2010/enablingjv.pdf, Retrieved January 8, 2006.
4. Mark W. Maier, "Architecting Principles for Systems-of-Systems." Proceeding of the 6th Annual INCOSE Symposium, pp. 567–574, 1996.
5. John Boardman, *et al.*, "Distinguishing Characteristics for a System of Systems," To be published. Internal communications 05-CSE01, Complex Systems Engineering Group, Stevens Institute of Technology, Hoboken, 2005.
6. United States Air Force, "United States Air Force Scientific Board: Report on System-of-Systems Engineering for Air force Capability Development," July 2005.
7. John Boardman, Mark A. Wilson, Allen Fairbairn, "Addressing the System of Systems Challenge," 15th Annual International INCOSE Symposium, 2005.
8. Edwin Morris, *et al.*, "Systems of Systems Interoperability (SOSI): Final Report," Carnegie Mellon Software Engineering Institute, April 2004.
9. Andreas Tolk, "Beyond Technical Interoperability — Introducing a Reference Model for Measures of Merit for Coalition Interoperability," 8th International Command and Control Research and Technology Symposium, June 2003.
10. Communications and Network Services, Glossary of Terms, University of California, www.net.berkeley.edu/dcns/glossary/, Retrieved March 6, 2006.
11. David A. Fisher and Howard F. Lipson, "Emergent Algorithms — A New Method for Enhancing Survivability in Unbounded Systems," Proceeding of the 32nd Hawaii International Conference on System Sciences, 1999.
12. Ozalp Babaoglu, *et al.*, "Design Patterns from Biology for Distributed Computing," Proceedings of the European Conference on Complex Systems, November 2005.
13. Robert W. Jacobs, "Model-Driven Development of Command and Control Capabilities For Joint and Coalition Warfare," Command and Control Research and Technology Symposium, June 2004.
14. Grace A. Lewis, Lutz Wrage, "Approaches to Constructive Interoperability," Carnegie Mellon Software Engineering Institute, December 2004.

15. J. T. Boardman, "Wholes and Parts — A Systems Approach," IEEE Transactions on Systems, Man, and Cybernetics, Vol. 25, No. 7, pp. 1150–1161, July 1995.
16. Richard Clarke, Janet Forbes, "JDEP Expansion in Joint Air and Missile Defense Mission Area and Beyond," 2003, ww.msiac.dmso.mil/sba_documents/JDEP%20JTAMD%20Events%20SIW%2003F.doc, Retrieved January 8, 2006.
17. Joint Interoperability Test Command, "JDEP Overview," October 2003, www.dtic.mil/ndia/2003systems/deb.ppt, Retrieved January 7, 2006.
18. John R. Clymer, "Role of Reductionism and Expansionism in System Design and Evaluation," INCOSE 1994.
19. Mary Linda Polydys, "Interoperability In DOD Acquisition Programs Through Enterprise "Architecting"." Acquisition Review Quarterly, Summer 2002, pp. 191–211.
20. Mary Maureen Brown and Rob Flowe, "Joint Capabilities and System-of-Systems Solutions," Defense Acquisition Review Journal, pp. 139–153.
21. Director, Operational Test and Evaluation, "Cooperative Engagement Capability (CEC)," FY98 Annual Report, 1998, www.fas.org/spp/starwars/program/dote98/98cec.htm, Retrieved January 8, 2006.
22. RADM Kate Paige (Retired), "Interoperability: The Navy View — An Acquisition Perspective," Naval Interoperability Workshop, May 30, 2001, www.dtic.mil/ndia/2001interop/paige.pdf, Retrieved January 8, 2006.
23. David Carney, *et al.*, "Topics in Interoperability: System-of-Systems Evolution," Carnegie Mellon Software Engineering Institute, March 2005.
24. Brian Sauser, *et al.*, "From TRL to SRL: The Concept of Systems Readiness Levels," Conference on Systems Engineering Research, April 2006.

Appendix E

APPLYING FRAMEWORKS TO MANAGE SoS ARCHITECTURE — ENGINEERING MANAGEMENT JOURNAL, DECEMBER 2008

MICHAEL DIMARIO
Sr. Program Manager
Lockheed Martin Corporation

ROBERT CLOUTIER
Research Associate Professor
School of Systems and Enterprises
Stevens Institute of Technology

DINESH VERMA
Dean and Professor
School of Systems and Enterprises
Stevens Institute of Technology

As systems become more complex, managing the development of these systems becomes more challenging. Additionally the integration of multiple independent systems increase the management challenge. System of Systems has unique attributes and which bring about unique management challenges. Architecture frameworks exist to organize and manage information about an architecture, and can be used to manage SoS architecture. One proposed approach was a framework that spans from component to enterprise, where enterprise is a SoS. A potential problem for this approach is that the lower half of the framework (from system down), is constructed using a reductionist approach while the upper part of these frameworks is constructed using a composition approach — that is, they are constructed by combining systems to make the broader enterprise system or SoS. So, does it stand to reason that the management approach for systems development should be the same as for SoS development? This paper addresses some of the reasons why they should be managed differently as well as offers a methodology for implementing the widely accepted Zachman Framework to facilitate the managing of SoS architectures.

Introduction

System of systems (SoS) comprises numerous constituent interdependent systems. These systems are autonomous and heterogeneous forming partnerships whereby their interoperability relationship produces capabilities or unintended consequences that do not originate from anyone individual constituent system. They exhibit the SoS characteristics of autonomy, belonging, connectivity, diversity, and emergence (Boardman and Sauser 2006). The interoperability relationships of the constituent systems create new behaviors of the holistic system. The picture that emerges is one of a holistic perspective versus a constituent perspective. The organization's systems engineering must form an enterprise architecture that supports the engineering of the SoS. The nature of a SoS has many disparate stakeholders representing their disparate systems as well as the capabilities to be derived for the SoS. The enterprise architecture must operate in support of the constituent systems as well as the formation of the SoS.

Many enterprise architecture frameworks have been reviewed to support a SoS environment because they help organize architecture descriptions and information as perceived by their users. We have chosen to extend and modify the Zachman Framework (ZF) (Zachman 1987) because it provides various artifacts needed to describe an information system as viewed by different stakeholder's perspectives of the system. Through the modification and extension of the Zachman Framework, a SoS engineering (SoSE) methodology emerges that enables the management of constituent systems and the SoS that is functionally dependent upon the SoS attributes.

Systems and System of Systems

There are numerous definitions of a system. INCOSE defines a system as an integrated set of elements that accomplish a defined objective. These elements include products (hardware, software, and firmware), processes, people, information, techniques, facilities, services, and other support elements (INCOSE 2006). Buede defines a system as "a set of components (subsystems, segments) acting together to achieve a set of common objectives via the accomplishment of a set of tasks" (Buede 2000). For the purposes of this paper, we will consider a system as something that has the ability to perform a set of tasks to satisfy a mission or objective. For instance, an automobile can move a person from one location to another and is a system. The motor with the automobile cannot perform a goal by itself.

Removed from the automobile and placed on the ground, it does nothing until it is combined with other parts of a system, e.g. a fuel delivery element, can it work in concert with those other parts to perform the goal of moving an individual to another location. This is not to minimize the complexity or importance of the motor. It is simply a subsystem of a broader system. The same holds true for the fuel delivery subsystem. It is an important part of the automobile, but is a subsystem in the automobile.

There are numerous definitions of SoS. INCOSE defines a SoS as "a system-of-interest whose system elements are themselves systems; typically these entail large scale inter-disciplinary problems with multiple, heterogeneous, distributed systems" (INCOSE 2006). The various definitions of systems do not Table 1: Differentiating a System from a System of Systems (Boardman and Sauser 2006) provide a detailed taxonomy to provide differentiation resulting in a dual perspective of a *system-of-interest is my element versus your system.* This perception extends to *your system is my SoS*, resulting in sufficient discussion on differentiating elements and systems (Simon 1962; Koestler 1967; Boardman 1995; Simon 1996).

Koestler (1967) managed this perceptual challenge with the introduction of *holons*, which are nodes on a hierarchic tree that behave partly as systems, or as elements, depending on how the observer perceives them. In this model there are intermediary forms of a series of levels in an ascending order of complexity. These levels are considered sub-systems which have characteristics of the system as well as its elements. This leads to the Janus Principle, which is the dichotomy of the systems and elements of autonomy and dependence (Koestler 1967). Within the concept of hierarchic order, members of the hierarchy have two faces looking in two directions, one looking up the hierarchy, and the other looking down the hierarchy. The *member as a system* is looking downward in descending order to less complex levels while the *member as an element* is looking upward in ascending order to more complex levels and is the dependent part of the hierarchy.

The system's spectrum displays a traditional ordinary system as a whole and its assemblies to a large SoS comprised of constituent systems. The attributes range from one of central control for a monolithic system to a mixture of attributes for a decentralized system to a set of attributes for a SoS. A primary assumption of a SoS is that they are open systems and are best understood in the context of their environments. It is also proposed that SoS have unique attributes unlike traditional ordinary systems (Maier 1996; Sage and Cuppan 2001; Keating, Rogers *et al.* 2003; DeLaurentis

Table 1. Differentiating a System from a System of Systems (Boardman and Sauser 2006).

Attribute	System	System of Systems
Autonomy	Autonomy is ceded by parts in order to grant autonomy to the system.	Autonomy is exercised by constituent systems in order to fulfill the purpose of the SoS.
Belonging	Parts are akin to family members; They did not choose themselves but came from parents. Belonging of parts is in their nature.	Constituent systems choose to belong on a cost/benefits basis; also in order to cause greater fulfillment of their own purposes, and because of belief in the SoS supra purpose.
Connectivity	Prescient design, along with parts, with high connectivity hidden in elements, and minimum connectivity among major subsystems.	Dynamically supplied by constituent systems with every possibility of myriad connections between constituent systems, possibly via a net-centric architecture, to enhance SoS capability.
Diversity	Managed i.e. reduced or minimized by modular hierarchy; parts' diversity encapsulated to create a known discrete module whose nature is to project simplicity into the next level of the hierarchy.	Increased diversity in SoS capability achieved by released autonomy, committed belonging, and open connectivity.
Emergence	Foreseen, both good and bad behavior, and designed in or tested out as appropriate.	Enhanced by deliberately not being foreseen, though its crucial importance is, and by creating an emergence capability climate, that will support early detection and elimination of bad behaviors.

and Crossley 2005; Saunders, Croom *et al.* 2005; Boardman and Sauser 2006). A set of attributes used for reference are shown in Table 1 as they relate to a system and a SoS (Boardman and Sauser 2006). A SoS is comprised of autonomous constituent systems fulfilling an objective as independent systems but are interdependent to fulfill holistic objectives of a SoS.

The constituent systems must perform an interoperation to fulfill a SoS capability in a collective manner whereby the capability does not exist on any of the constituent systems. SoS behavior and capabilities are that of the whole and not properties of the constituent systems (elements) or that can be deduced by properties of the elements. The emergent behavior is a product of the interactions, not the sum of the actions of the constituent elements. Behavior arises from the organization of the elements without having to identify the properties of the individual elements. For example, a computer is an organization of elementary functional components. Only the function performed by those components is relevant to the behavior of the whole system (Simon 1996).

Architecture and Architecture Frameworks

While architecture is an overused and misunderstood word in the engineering lexicon, it is necessary to describe the structure of a system. Every system has an architecture, whether it is explicitly or implicitly designed and documented. There are many definitions for architecture. INCOSE defines system architecture as "The arrangement of elements and subsystems and the allocation of functions to them to meet system requirements" (INCOSE 2006).

IEEE 1471 defines architecture as the "fundamental organization of a system embodied in its components, their relationships to each other, and to the environment, and the principles guiding its design and evolution" (Institute of Electrical and Electronics Engineers 2000). While the words between the two definitions are somewhat different, let us agree that the organizing concept behind architecture is an organizing of a system, into constituent parts as specified through requirements, to satisfy a desired goal.

One challenge when discussing architecture is to understand what part of the architecture is under discussion. Frameworks help in the organizing architectural information, as the elements and subsystems, seen by different consumers of that information while providing a technique for thinking about architecture and organizing architecture descriptions. The ZF, shown in Table 2, provides a sample of what type of information might be found in each of the intersections between the perspectives (rows) and the aspects (columns).

The ZF identifies the various artifacts needed to describe an information system, as viewed by different stakeholder's perspectives of the system. "The framework is a classification schema for descriptive representations of an enterprise. By observing design artifacts of physical objects like

Table 2. Zachman Framework™ for Enterprise Architecture (Zachman 1987).

	What Data	How Function	Where Network	Who People	When Time	Why Motivation
Scope [contextual] Planner	List of important things to the Business	List of processes the Business performs	List of locations in which the Business operates	List of organizations important to the Business	List of events/cycles significant to the business	List of Business goals/strategies
Business Model [conceptual] Owner	Semantic model	Business process model	Business logistics system	Work flow model	Master schedule	Business plan
System Model [logical] Designer	Logical data model	Application architecture	Distributed system architecture	Human interface architecture	Processing structure	Business rule model
Technology Model [physical] Builder	Physical data model	System design	Technology architecture	Presentation architecture	Control structure	Rule design
Detailed Presentation [out-of-context] Subcontractor	Data definition	Computer Program	Network architecture	Security architecture	Timing definition	Rule specification
Functioning Enterprise	e.g.: Data	e.g.: Function	e.g.: Network	e.g.: Organization	e.g.: Schedule	e.g.: Strategy

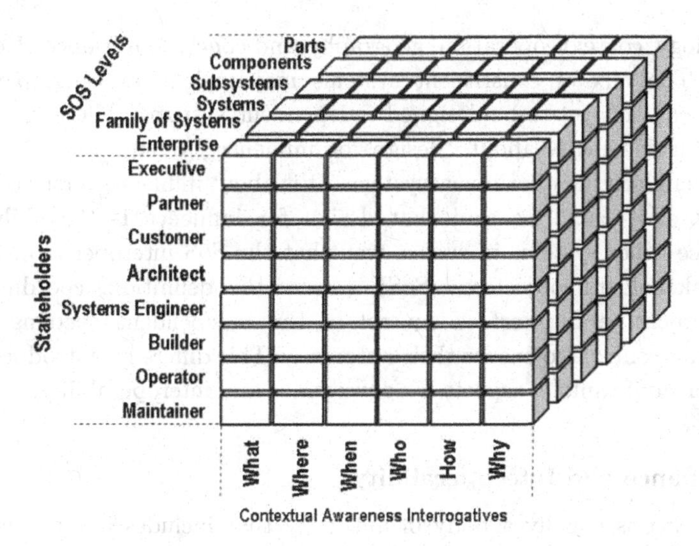

Fig. 1. Three Dimensional Architectural Framework (Morganwalp and Sage 2002/2003).

airplanes, buildings, ships, and computers, John Zachman derived the framework" (O'Rourke, Fishman *et al.* 2003). The artifacts captured in the ZF can be viewed either by the perspective of the person viewing the artifact or the aspect of the content or subject focus. Over time, Zachman has added additional information to his framework. Figure 1 represents the current state of the ZF.

Initially intended for use by Information Technology organizations, the ZF has since been applied to facilitate understanding of information architectures in multiple domains. The ZF provides a contextual basis for discussions among stakeholders with disparate levels of understanding regarding the enterprise under discussion in a holistic manner (Morganwalp and Sage 2004). Based on the organizing benefits of frameworks, and their use to manage aspects of complex enterprise architecture; it follows to investigate a framework approach to discussing and managing SoS. The rest of the paper will propose such a framework.

Systems of Systems Engineering Management

SoS Engineering (SoSE) is described as "The design, development, operation, and transformation of metasystems that must function as an integrated complex system to produce desirable results" (Keating, Rogers *et al.* 2003). A metasystem is a SoS "comprised of multiple embedded and interrelated autonomous complex subsystems that can be diverse in technology, context, operation, geography, and conceptual frame" (Keating 2005). The mix of constituent systems may include existing, partially developed, and yet-to-be-designed independent systems. The engineering and management is about "design for influence" versus point or single system engineering of ordinary systems. It is about influence versus control. Therefore, it could be said that design for influence is the ability to influence other systems in such a way that the SoS interoperation fulfills the stakeholder's objectives. SoSE fosters the definition, coordination, development, and interface control of the independent systems while providing controls to ensure their autonomy. This difference introduces the need for SoSE unique aspects for governance and interoperability.

Governance and Interoperability

Governance is usually a body of authority that includes but not limited to structures of authority and collaboration to allocate resources and coordinate management control activity. Governance is also the process and systems by which a SoS is managed and built. For a single system the majority of stakeholders manages only a single system and has very specific roles in systems engineering. This governance is hierarchical with a lead engineer and program manager. For SoS, the stakeholders tend to be more diverse because they have interest in other systems as well. In addition, the SoS involves other stakeholders as "unanticipated users" as the SoS is more dynamic and its diversity becomes more apparent as it matures.

Governance is a primary problem in managing, developing, and deploying SoS because most, if not all current governance structures are at the individual constituent system levels. This is because the structure is focused upon the individual systems due to the, in reference to Department of Defense (DoD), Work Breakdown Structure (WBS) as it is designed to support incremental acquisition of operational capabilities (Wang, Lane *et al.* 2006) and promotes management and engineering in a reductionist manner promoting individual systems and components versus a systems composition manner that promotes SoS. The SoSE and SoS acquisition is new to the DoD community and requires new perspectives and training in transforming existing contract and program structures to a SoS acquisition model.

Interoperability has many domain-oriented definitions. A definition used in this context is "the ability of a collection of communicating entities to share specified information and operate on that information according to a shared operational semantics to achieve a specified purpose in a given context" (Carney, Fisher *et al.* 2005). Systems interoperability or interoperation assumes that systems have a relationship and that a system provides a service to another. In other words, in a SoS context there are cooperative interactions among loosely autonomous systems to adaptively fulfill a system-wide purpose. The constituent systems of a SoS must communicate in some form otherwise it is not realizable (DiMario 2006). Syntactic communication is the initial step of interoperability whereby a connection is established and semantic communication is necessary for the communicating systems to know what to do with the information that is communicated adding value to the SoS.

System of Systems Framework

A goal of SoS development is to place an emphasis on how systems interact to accomplish their common collective purpose with a de-emphasis on the individual systems. The continuing development and independence of the constituent systems as well as the common collective purpose is difficult to achieve solely through the selection of correct interfaces (Kuras and White 2005). In addition, changes and evolution of the constituent systems are difficult to align individual system priorities with the SoS holistic priorities due to the lack of governance. Even though the constituent systems may continue to connect, their ability to transfer and make use of the information may be hindered or result in unanticipated emergent behavior. The difficulty arises due to the disparate management of all the independent systems. Decision making and control is distributed without governance.

The ZF provides the flexibility of organizing the necessary information to disparate stakeholders of an enterprise of mutual interest. This flexibility readily supports an extension of a three dimensional perspective for a SoS by adding the SoS attributes as a third dimension. A three dimensional SoS architecture framework employing the Zachman architectural framework is not new; an example is shown in Figure 2 whereby the various hierarchical SoS levels are mapped (Morganwalp and Sage 2002/2003). This graphical representation represents the systems engineering life cycles focused on the stakeholder views of engineering a system, contextual awareness interrogatives, and the SoS levels.

The essence of a SoS is constituted by the five attributes shown in Table 1. The SoS enterprise architecture should be engineered with

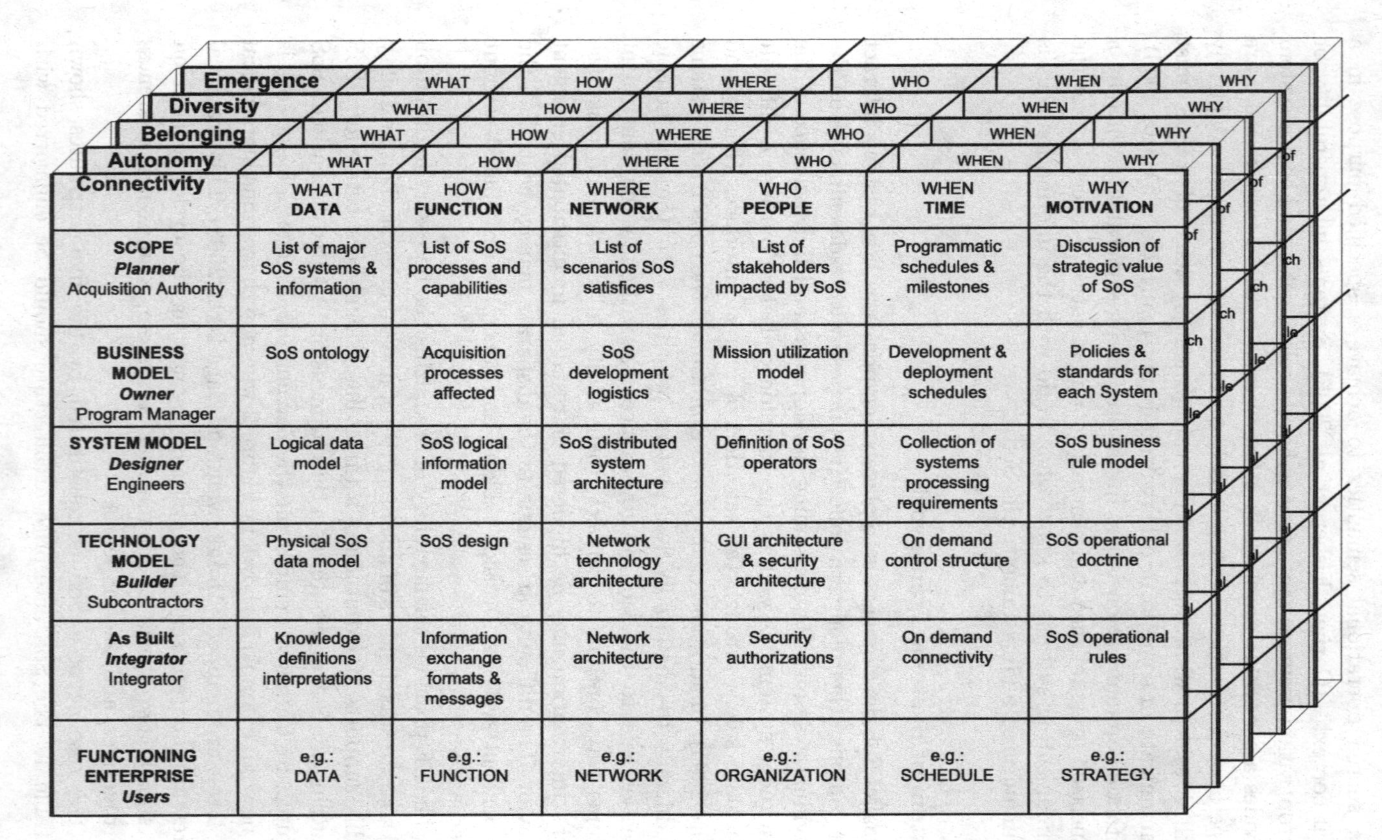

Fig. 2. SoS Connectivity Attribute Information Architecture Using Zachman Framework.

consideration of the SoS attributes to operate as a whole. The constituent systems of the SoS must operate independently and are generally managed independently of one another contributing to the whole when required. There is a need for a suitable architectural framework to support the SoS environment of systems that adapt, evolve, and possess emergent behavior. Systems engineering is the management of technology that controls the total life cycle process that meets user needs and a suitable framework to enable the engineering of a system.

The SoS attributes may be managed by mapping them to the Zachman architectural framework of stakeholders that share interests and concerns and express their interests and concerns via the contextual awareness interrogatives as shown in Figure 2. The cell intersection of the stakeholders — Acquisition Authority, Program Manager, Engineers, Subcontractors, and Integrator and the Contextual Awareness Interrogatives — What, How, Where, Who, When, Who, and Why for each of the SoS attributes provides a unique perspective and abstraction. Through creating a set of unique perspectives and abstractions for each attribute, the SoS may be developed, deployed, and managed in a design for influence manner achieving holistic capabilities and in a manner that maintains constituent system management and independence. The goal of performing SoSE in a compositional manner may be achievable aligning the priorities of a SoS as well as the individual systems.

The SoS attribute connectivity is chosen to show the unique abstractions for a SoS attribute and will most likely be very different for the other attributes, even though the stakeholders remain the same. Connectivity was chosen for this example since a SoS cannot exist if the constituent systems cannot connect and exchange information. The SoS connectivity attribute is about interoperation among constituent systems and must support the flow of data and knowledge syntactically and semantically. The abstractions shown in the individual cells indicate actions or artifacts required to achieve connectivity in the SoS environment aligning SoS and individual system priorities.

Conclusions

This paper describes the nature of a SoS and the application of the ZF in defining the enterprise architecture for SoSE. The ZF was modified and extended using the SoS attributes as functions for the stakeholders and contextual awareness interrogatives. The important points presented in this paper include:

- ZF lends itself easily for adaptation and SoS usage
- SoS attributes provide focus for the stakeholders
- Artifacts for SoSE and management are identified for the connectivity attribute
- Abstractions and perspectives per attribute provide SoS composability
- Constituent system and SoS priorities are adjudicated by using same stakeholders
- The problems introduced due to the absence SoS governance may be adjudicated through the interaction of the attribute cell results of interrogatives and stakeholders
- SoS Design for Influence is enabled through the modified Zachman Framework

Recommendations for Future Research

Future research and investigation is required to extend this methodology to the other attributes. The collective interactions of the artifacts contained in any single ZF cell on a single plane are complex. The addition of multiple planes, or dimensions, to the ZF to represent the unique SoS attributes compounds the complexity of those interactions. The effects of this complexity on the ZF require further investigation. Validation of the cells in real projects is required to fully test this methodology. Through validation will provide sensitivity of the attributes in a SoS environment and may reveal improved processes to manage the priorities of individuals systems and the SoS creating processes for SoSE and management.

References

1. Boardman, J. (1995). "Wholes and Parts — A Systems Approach." *IEEE Transactions on Systems, Man and Cybernetics* **25**(7): 1150–1161.
2. Boardman, J. and B. Sauser (2006). *System of Systems — the meaning of of.* IEEE International Conference on System of Systems Engineering, Los Angeles, CA.
3. Buede, D. M. (2000). *The Engineering Design Of Systems: Models And Methods.* New York, John Wiley & Sons, Inc.
4. Carney, D., D. Fisher, *et al.* (2005). Some Current Approaches to Interoperability. Pittsburgh, Software Engineering Institute: 27.
5. DeLaurentis, D. A. and W. A. Crossley (2005). *A Taxonomy-based Perspective for Systems of Systems Design Methods.* IEEE International Conference on Systems, Man and Cybernetics.
6. DiMario, M. J. (2006). *System of Systems Interoperability Types and Characteristics in Joint Command and Control.* IEEE Conference on System of Systems Engineering, Los Angeles, CA.

7. INCOSE (2006). *INCOSE Systems Engineering Handbook V.3*, INCOSE.
8. Institute of Electrical and Electronics Engineers (2000). *IEEE Std 1471-2000: IEEE Recommended Practice for Architectural Description of Software-Intensive Systems.* Piscataway, IEEE.
9. Keating, C., R. Rogers, *et al.* (2003). "System of Systems Engineering." *Engineering Management Journal* **15**(3): 36–45.
10. Keating, C. B. (2005). *Research Foundations for System of Systems Engineering.* 2005 IEEE International Conference on Systems, Man and Cybernetics, Waikoloa, Hawaii.
11. Koestler, A. (1967). *The Ghost In The Machine.* New York, The Macmillan Company.
12. Kuras, M. L. and B. E. White (2005). *Engineering Enterprises Using Complex-System Engineering.* INCOSE.
13. Maier, M. W. (1996). *Architecting Principles for Systems-of-Systems.* Proceeding of the 6th Annual INCOSE Symposium.
14. Morganwalp, J. and A. P. Sage (2002/2003). "A System of Systems Focused Enterprise Architecture Framework." *Information Knowledge Systems Management* **3**(2): 87–105.
15. Morganwalp, J. and A. P. Sage (2004). "Enterprise Architecture Measures of Effectiveness." *International Journal of Technology, Policy and Management* **4**(1): 81–94.
16. O'Rourke, C., N. Fishman, *et al.* (2003). Enterprise Architecture Using the Zachman Framework. Boston, Thomson Course Technology: 33.
17. Sage, A. P. and C. D. Cuppan (2001). "On the Systems Engineering and Management of Systems of Systems and Federation of Systems." *Information Knowledge Systems Management* **2**(4): 325–345.
18. Saunders, T., C. Croom, *et al.* (2005). System-of-Systems Engineering for Air Force Capability Development: Executive Summary and Annotated Brief.
19. Simon, H. A. (1962). "The Architecture of Complexity." *Proceedings American Philosophical Society* **106**(6): 467–482.
20. Simon, H. A. (1996). *The Sciences of the Artificial.* Cambridge, The MIT Press.
21. Wang, G., J. A. Lane, *et al.* (2006). *Towards a Work Breakdown Structure for Net Centric System of Systems Engineering and Management.* INCOSE 2006 — 16th Annual International Symposium Proceedings, System Engineering: Shining Light on the Tough Issues, Orlando, FL.
22. Zachman, J. A. (1987). "Introduction to Enterprise Architecture." *IBM Systems Journal* 26(3).

Appendix F

"SATISFICING" SYSTEM OF SYSTEMS (SoS) USING DYNAMIC AND STATIC DOCTRINES — SUBMITTED FOR PUBLICATION

ALEX GOROD
Systomics Laboratory, School of Systems and Enterprises
Stevens Institute of Technology, Hoboken, NJ 07030, USA
agorodis@stevens.edu

MICHAEL DIMARIO
Lockheed Martin MS2, Moorestown, NJ 08057, USA
michael.j.dimario@lmco.com

BRIAN SAUSER* and JOHN BOARDMAN
Systomics Laboratory, School of Systems and Enterprises
Stevens Institute of Technology, Hoboken, NJ 07030, USA
* *bsauser@stevens.edu* [§] *boardman@stevens.edu*

System of Systems (SoS) is a concept used to describe capabilities, resulting from the network-centric connectivity of systems whereby the elements of a system are classified as *summative* or *constitutive*. The summative system owes its capabilities to the summation of the characteristics and sub-capabilities of its elements. The constitutive system has capabilities greater than the summation of its elements. The constitutive system concept demonstrates a potential need to shift the traditional systems engineering focus to address the problems of multiple integrated complex systems for which a final solution is not necessarily the end expectation. The objective involves "satisficing" a holistic system. By examining the Auto Battle Management Aid (ABMA) case study, we conclude that dynamic and static doctrines provide us with an opportunity to observe a newly developed management framework. The goal includes creating a framework that would allow us to effectively "satisfice" the management demands of an ever-changing SoS environment.

Keywords: System of Systems Engineering Management Framework; System of Systems; SoS; System of Systems Engineering; SoSE; Satisficing; Dynamic Doctrine; Static Doctrine; SoS Characteristics.

Biographical Notes

Alex Gorod received a B.S. in Information Systems and a M.S. in Telecommunications from Pace University. Prior to his graduate studies, he held a Research Analyst position at Salomon Smith Barney. He is currently a Robert Crooks Stanley Doctoral Fellow in Engineering Management at Stevens Institute of Technology, with research interests in the area of management of complex systems. He is also the President of the Stevens Student Chapter of the International Council on Systems Engineering (INCOSE).

Michael DiMario received a B.S. in Natural Science and a M.S in Computer Science from the University of Wisconsin and a M.B.A. from the Illinois Institute of Technology. He is a senior program manager with Lockheed Martin managing science and technology programs dealing with complex systems. Prior to joining Lockheed Martin, he served various leadership-engineering roles at Lucent Bell Laboratories. He is a Doctoral Candidate at Stevens Institute of Technology with research interests in cooperative complex systems and is a founding member of the IEEE Systems of Systems Consortium.

Brian Sauser holds a B.S. from Texas A&M University, a M.S. from Rutgers, The State University of New Jersey, and a Ph.D. from Stevens Institute of Technology. He is currently an Assistant Professor in the School of Systems and Enterprises at Stevens Institute of Technology. His research interests are in theories, tools, and methods for bridging the gap between systems engineering and project management for managing complex systems. This includes the advancement of systems theory in the pursuit of a Biology of Systems, system and enterprise maturity assessment for system and enterprise management, and systems engineering capability assessment.

John Boardman graduated with 1st Class Honors in Electrical Engineering from the University of Liverpool, from where he also obtained his PhD. He is currently a Distinguished Service Professor in the School of Systems and Enterprises at Stevens Institute of Technology. Before coming to Stevens Institute of Technology he held positions at the University of Portsmouth as the GEC Marconi Professor of Systems Engineering and Director of the School of Systems Engineering and later Dean of the College of Technology. He is a Chartered Engineer and Fellow of the Institute of Engineering and Technology and the International Council on Systems Engineering (INCOSE).

Introduction

System of systems (SoS) has emerged to be a concept used to describe capabilities resulting from the network-centric connectivity of systems (Balci, 2008) and is generally associated with large-scale inter-disciplinary objectives with multiple, heterogeneous, and distributed systems at various levels and domains. Other terms and concepts have appeared such as Family of Systems (FoS) (U.S. Joint Chiefs of Staff, 2005), and Federation of Systems (F-SoS) (Krygiel, 1999) as well as greater than forty variant definitions of SoS (Boardman and Sauser, 2006) and their inconsistencies (Saunders, Croom *et al.*, 2005). SoS naturally exists where there are benefits derived from systems interoperating with other systems such as in a large transportation system (DeLaurentis, Lewe *et al.*, 2003; Sauser, Boardman *et al.*, 2008), space exploration (Aldridge, 2004), and healthcare (Reid, Compton *et al.*, 2005).

SoS comprise numerous constituent interdependent systems and are an integration of complex metasystems (Keating, Rogers *et al.*, 2003). This description suggests that the systems are autonomous and heterogeneous. They form partnerships whereby their interoperability relationship produces capabilities or unintended consequences because of emergent behavior that does not originate from any single constituent system. Additionally, these capabilities and consequences cannot be deduced by properties of any collective constituent systems. Therefore, the interoperability relationships of the constituent systems create new behaviors of the holistic system.

The holistic concept of systems is not new. It was introduced in General Systems Theory (GST) as a science of wholeness whereby general systems laws apply to any system of a certain type independent of its properties. The elements of a system are classified into two types: *summative* and *constitutive.* The summative system owes its capabilities to the summation of the characteristics and sub-capabilities of its elements. The properties of the elements are universally the same in all environments and have no relationships. The constitutive system has capabilities that are greater than the summation of its elements because the effects are dependent upon the context of their properties and their relationships (Bertalanffy, 1969). Boulding described this concept as a SoS which may perform the function of *gestalt,* a pattern so unified as a whole that its properties cannot be derived from its parts (Boulding, 1956).

Traditionally, engineered systems or executed system engineering (SE) focus on a system with a well-defined problem set, known boundaries, and specified system life cycle timeframe. Typically, the objective involves optimizing a system using an explicit process-oriented approach. However, the constitutive SoS concept shifts the traditional SE focus to address the problems of multiple integrated complex systems (System of Systems Engineering (SoSE). Further, a final solution is not necessarily the end expectation (Keating, Rogers *et al.*, 2003; Gorod, Sauser *et al.*, 2008) but may be composeable according to a stakeholder's system's perspective (DiMario, Cloutier *et al.*, 2008). Consequently, the objective involves a "satisficing" process wherein a holistic system employs a methodology-oriented approach because of dynamic boundaries, an emergent problem set and pluralistic goals (Gorod, Sauser *et al.*, 2008b). This paper will further explore the "satisficing" concept for SoS as described by Gorod, Sauser *et al.* (2008b) and propose a framework that will allow "satisficing" a SoS using dynamic and static doctrine.

"Satisficing" SoS

When we make a decision, the natural desire is to find the "best" solution to the problem. Unfortunately, real life environment imposes significant constraints that limit our decision making process. Thus, we are frequently settling for satisfactory rather than ideal goals because the constraints are too costly. Byron (1998) described this fact in the following manner: "A natural demand is that instrumentally rational actions implement the *best* means to one's given ends. Optimizing conceptions of rationality endorse this demand. A competing conception of rationality—the satisficing conception—weakens this requirement and permits some rational actions to implement (merely) *satisfactory* means to the agent's given ends" (Byron, 1998). For perspective, the following is a list of some of the constraints that have been described for a decision making process (Zsambok and Klein, 1997; Berryman, 2006):

- Ill-structured problems;
- Uncertain and dynamic environments;
- Shifting, poorly defined or competing goals;
- Action, feedback loops;
- Time stress;
- High stakes;
- Multiple players;
- Organizational goals and norms.

Thus, optimizing involves finding the best option for these constraints based on an individual's goal (Pettit, 1984). According to Pettit (1984) and Byron (1998), optimization consists of the following three steps:

1. Itemize all available options.
2. Assess each one of them.
3. Select the best.

Unfortunately, there rarely exists an absolute optimal solution when we are faced with a complex dynamic environment. Therefore, a Pareto-optimal solution is often used. Kiyota, Tsuji *et al.* (2003) even described the Pareto-optimal solutions as, "... a satisficing one for the decision maker." (Kiyota, Tsuji *et al.*, 2003).

The term "satisficing," introduced by Noble Prize Laureate Herbert A. Simon, indicates that in a complex dynamic environment true optimization or the "best" result cannot always be achieved. Thus, he suggested, the use of a "good enough" approach (Simon, 1955; Simon, 1956). Schmidtz (1996) further explains that, "Satisficing emerges not as an alternative to optimizing as a model of rationality but rather as an alternative to local optimizing as a strategy for pursuing global optima" (Schmidtz, 1996). Satisficing, in contrast to Optimizing, involves the following steps (Pettit, 1984; Byron, 1998):

1. Determine a benchmark at which any option that meets or exceeds it becomes "good enough."
2. Begin to itemize and assess all available options.
3. Select the first option that is good enough in the benchmark context.

We can deduct from this comparison that satisficing and optimization involve different decision-making structures. In the case of satisficing, the satisficer will set a benchmark, start itemizing, assess the options, and finally select the first one that is greater than or equal to the benchmark value. Conversely, the optimizer will itemize and assess all options and then select the best one. In Figure 1, Schmidtz illustrated two stopping rules for both satisficing and optimizing within a "local utility space." The stopping rule for satisficing is when one of the options meets or exceeds the "satisficer's aspiration level" (Schmidtz, 1996). While the optimizer considers options until they reach the "vertical constraint" on the number of options that can be considered (Byron, 1998).

We contend that the satisficing approach can be applied to SoS because it may not be feasible to optimize a constitutive SoS due to the uncertainty

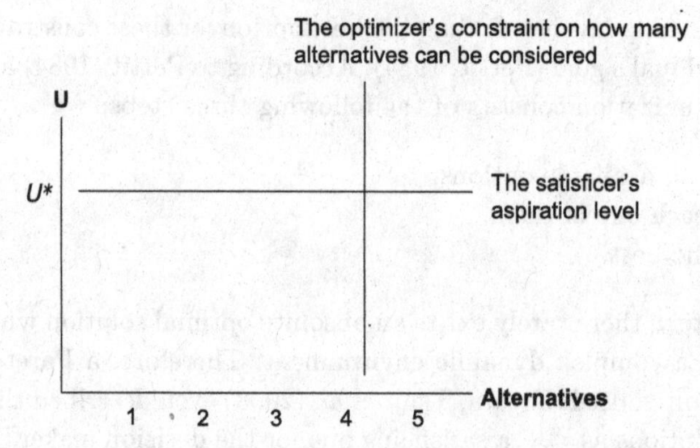

Fig. 1. Two Stopping Rules of Satisficing and Optimality Contrasted (Schmidtz, 1996).

and complexity of its nature. Therefore, the best management solution would be to "satisfice" a SoS.

Static and Dynamic Doctrines

A doctrine is frequently viewed as a belief system made of a set of rules. A general definition is "something taught" and "a principle or body of principles presented by a specific field, system, or organization for acceptance or belief" (Webster's Dictionary, 2001). Doctrine typically is executed within an environment that is subject to change due to changes in technology, social values, and the general principles and rules of the doctrine. A static doctrine is born from knowledge and becomes a static belief system. A dynamic doctrine, however, is seemingly contradictory as it can change from a static state.

A static doctrine not founded in change means the system is reactive or has a delayed reaction, described as a post factum act. A dynamic doctrine, however, is proactive and refers to preset conditions prior to the actual event. An analogy could be a thermostat. It is used to preset temperature to a desired degree. Then, it dynamically adjusts temperature to the preset level, or a change in the existing doctrine or a new state bestowing a dynamic doctrine. At the same time, it allows a user to intervene and change the temperature statically using the reactive method bestowing a static doctrine. A dynamic environment is continuously asking the system what it wants to do. A static environment does not require the system to review the states of the environment while it is deciding on an action. The same mechanism can be applied to the SoSE domain in the way that it utilizes both the dynamic and static doctrine to effectively manage a SoS.

Dynamic Doctrine and SoS Characteristics

The law of requisite variety suggests that for a system to survive in its environment the variety of choices the system may have is at least equal to the variety of influences that the environment imposes. Thus, a dynamic doctrine must be able to accommodate environmental impositions in order to support survival. In the case of our thermostat, the thermostat must regulate the activities of the environmental system with at least the same number of degrees of freedom as the system it is trying to regulate. In a constitutive SoS, this becomes a complex variation of interactions of systems with their own doctrines supporting a grander doctrine. The five distinguishing characteristics of SoS as defined by Boardman-Sauser (2006) are used to represent the dynamic doctrine. The characteristics are:

- ***Autonomy***: The ability of a system as part of SoS to make independent choices. This includes managerial and operational independence while accomplishing the purpose of SoS.
- ***Belonging***: Constituent systems have the right and ability to choose to belong to SoS. The choice is based on their own needs/believes and/or fulfillment.
- ***Connectivity***: The ability to stay connected to other constituent systems.
- ***Diversity***: Evidence of visible heterogeneity; "Increased diversity in SoS capability achieved by released autonomy, committed belonging, and open connectivity" (Boardman and Sauser, 2006).
- ***Emergence***: Formation of new properties because of developmental or an evolutionary process.

These characteristics were chosen because a characterization approach to understanding a SoS has been effective (Gorod, Sauser *et al.*, 2008) and Boardman and Sauser proposed that these characteristics "are influenced by fluxes in realizing or recognizing a system" (Sauser and Boardman, 2006). Thus, it allows for a dynamic nature in the characterization of a SoS.

Figure 2 depicts the spectrum of a system classification using the Boardman-Sauser Characteristics with their fluxes (or paradoxes). The left side of the figure exemplifies attributes distinctive to an Assembly type system (i.e. system of components), and the right side portrays the SoS.

Assembly
(e.g: CD player or cell phone)

System of Systems
(e.g: Internet, IDS or FCS)

Conformance
Autonomy is ceded by parts in order to grant autonomy to the whole

Autonomy

Independence
Autonomy is exercised by constituent systems in order to fulfill the purpose of the SoS

Centralization
Parts are akin to family members; they did not choose themselves but came from parents. Belonging of parts is in their nature.

Belonging

Decentralization
Constituent systems choose to belong on a cost/benefits basis; also in order to cause greater fulfillment of their own purposes, and because of belief in the SoS supra purpose

Platform-Centric
Prescient design, along with parts, with high connectivity hidden in elements, and minimum connectivity among major parts

Connectivity

Network-Centric
Dynamically supplied by constituent systems with every possibility of myriad connections between constituent systems, possibly via a net-centric architecture, to enhance SoS capability

Homogeneous
Managed i.e. reduced or minimized by modular hierarchy; parts' diversity encapsulated to create a known discrete module whose nature is to project simplicity into next level of the hierarchy

Diversity

Heterogeneous
Increased diversity in SoS capability achieved by released autonomy, committed belonging, and open connectivity

Foreseen
Foreseen, both good and bad behavior, and designed in or tested out as appropriate

Emergence

Indeterminable
Enhanced by deliberately not being foreseen, though its crucial importance is, and by creating and emergence capability climate, that will support early detection and elimination of bad behaviors

Fig. 2. Systems characterization, adopted from (Gorod, Gandhi *et al.*, 2008).

Static Doctrine and SoSE Management Conceptual Areas

A static doctrine that cannot be modified runs the risk of becoming dogmatic. A static doctrine is based on principles that do not represent its environment, and these principles can force rules that do not comply within this environment. As we described earlier, static doctrine allows us to reactively intervene and influence certain characteristics of a SoS that, as a result, will change the actual state of a SoS. To accomplish this, Gorod, Gove *et al.*, (2007) proposed to utilize modified Fault, Configuration, Accounting, Performance, and Security (FCAPS) network principles (SoSE Management Conceptual Areas). The SoSE Management Conceptual Areas can be used as the individual triggers to govern SoS (Gorod, Sauser *et al.*, 2008).

The SoSE Management Conceptual Areas can be described in the following manner (Gorod, Gove *et al.*, 2007):

Risk Management: Monitor, identify, asses, analyze, and mitigate risk encountered in the SoS

Configuration Management: Direct and coordinate through functioning and software management

Performance Management: Monitor and measure performance of SoS for it to be maintained at an appropriate level

Policy Management: Provide SoS access to authorized processes and protect SoS from illegal access

Resource Management: Coordinate and allocate SoS assets based on usage and utilization information of Systems' in the SoS

Framework for "Satisficing" SoS

The framework is described in terms of dynamic and static doctrine feedback loops. In a dynamic loop, the system allows fast automatic adjustments of the preset conditions. The static loop provides us with an opportunity to intervene and make reactive changes in achieving "satisficing" goals. Figure 3 serves as an illustration of these processes and indicates the flow of information through the two loops. An analogous depiction can be found in the SoSE Management Framework proposed by Gorod, Sauser *et al.* (2008). Overall, Figures 4 and 5 provide "the final result of interaction of different processes within SoS and the functionality of the effective SoSE management framework" (Gorod, Sauser *et al.* 2008).

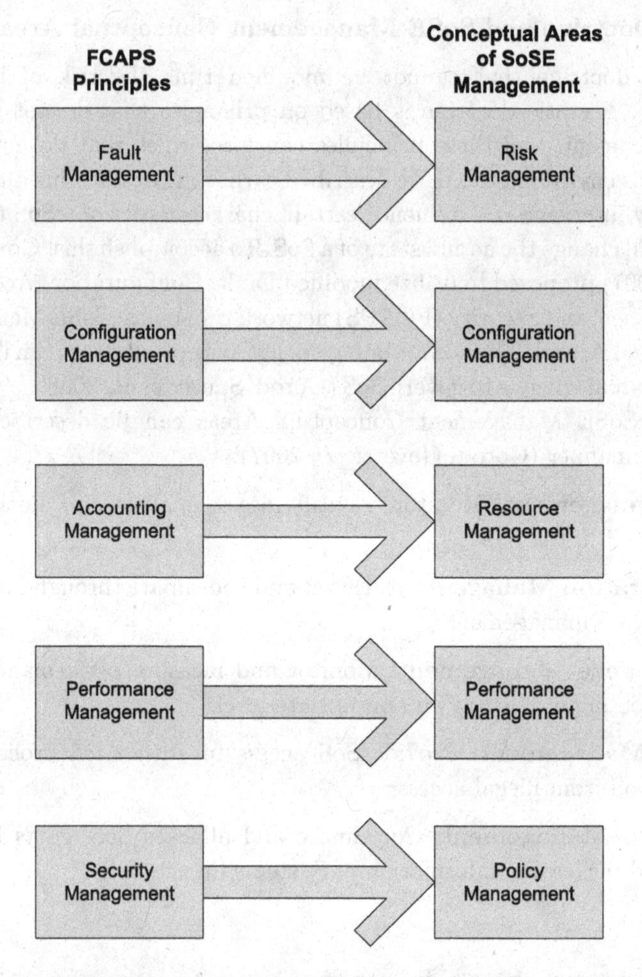

Fig. 3. SoSE Management Conceptual Areas (Gorod, Gove *et al.*, 2007).

Case Study of Auto Battle Management Aids

A case study of a concept of operations for an Auto Battle Management Aid (ABMA) exemplifies the complexities of static and dynamic SoS as well as the real time management and doctrine of such a system. Battle managers were recognized early in the development of the Strategic Defense Initiative Ballistic Missile Defense (SDI BMD) system of the 1980s. Battle management is defined as "The management of activities within the operational environment based on the commands, direction, and guidance given by appropriate

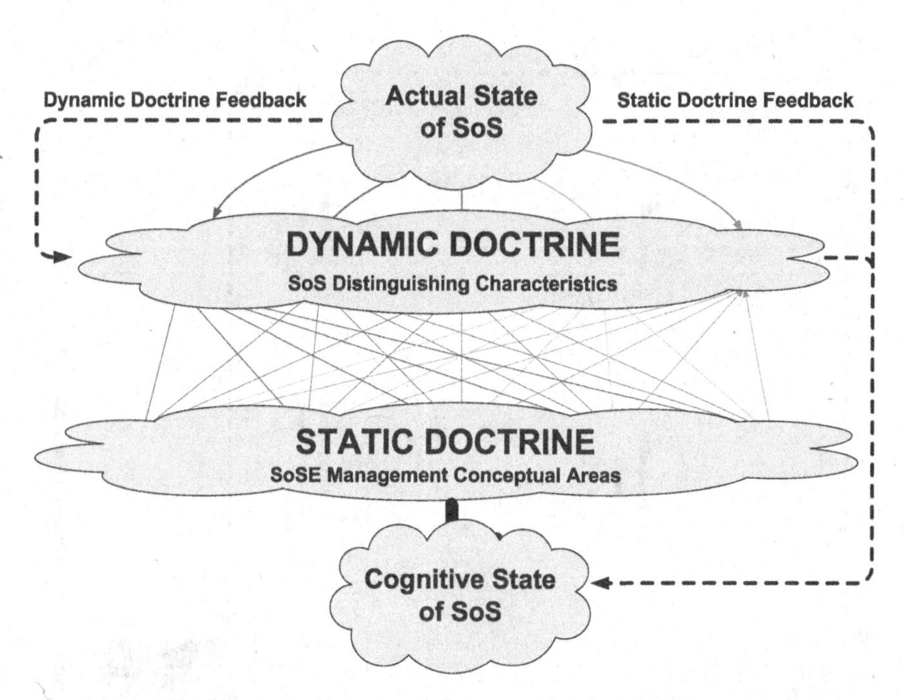

Fig. 4. Framework for "Satisficing" SoS using Dynamic and Static Doctrine.

authority" (U.S. Department of Defense, 2001). An SDI BMD ABM was required to reduce time, cognitive load, and to manage autonomous computer systems (U.S. Congress, 1988). The cognitive load is associated with the networking of systems logic, information processing, information exchange, and services as well as the semantic interpretation and translation of the exchanged data among the systems. Autonomic computing is required to reduce the costs of increased complexity and the ability to quickly respond to a large and asymmetric threat-enriched environment. It also mitigates the need for operational interaction by automating the response of the network event or actions based on a centralized doctrine. The autonomic response recognizes different classes of events and their thresholds for appropriate response. The automaticity is based on the tempo of various levels required for a response. For example, for the US Navy the tempo may range from months to weeks at command centers based on land and days to hours to seconds within a Navy battle group.

The objective of an ABMA and value to the DoD is to reduce cognitive load by reducing the decision time required to respond in a rapidly evolving

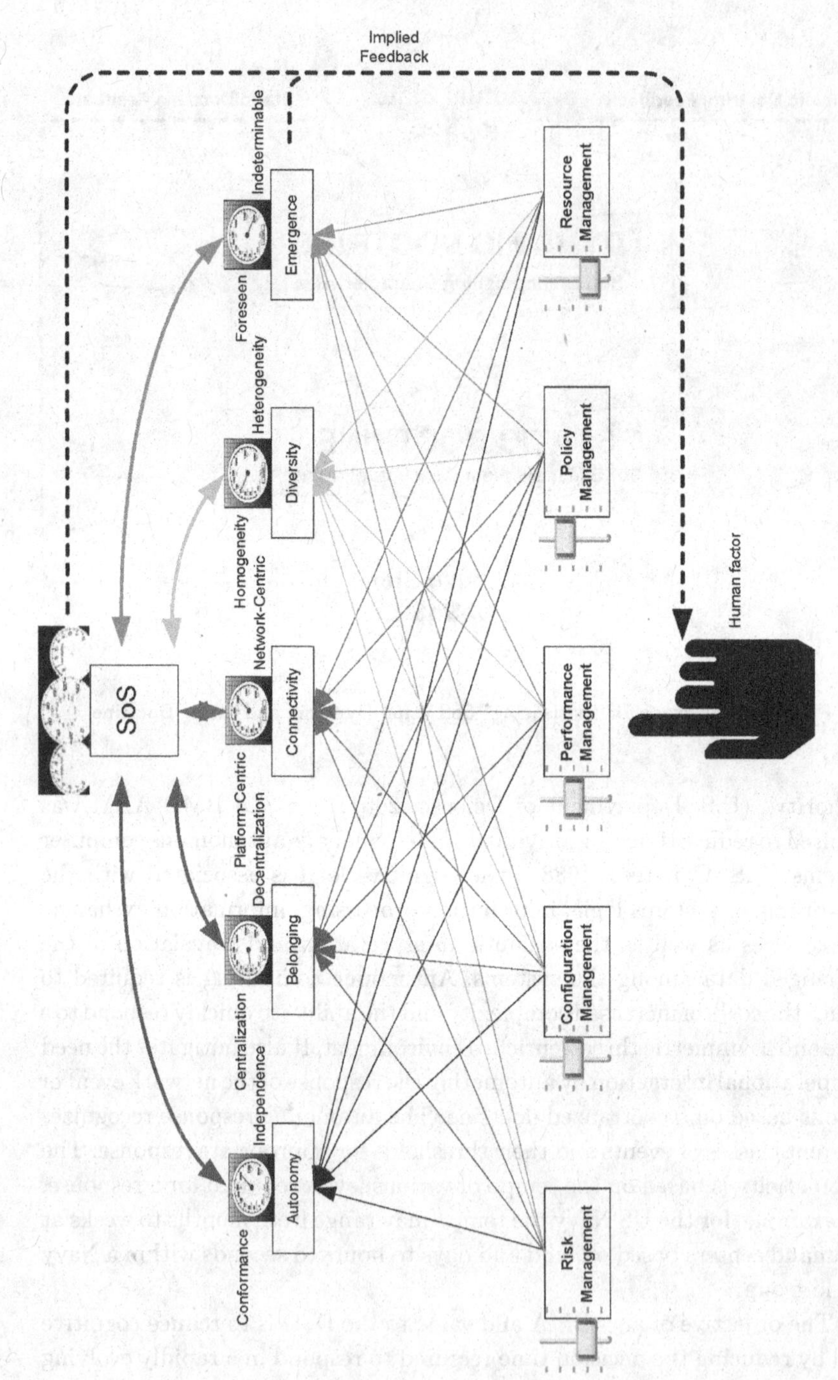

Fig. 5. SoSE Management Framework (Gorod, Sauser *et al.*, 2008).

dynamic threat environment. The complexities of early 21st century warfare has significantly changed from the conflicts of the 20th century (U.S. Joint Chiefs of Staff, 2000). An ABMA becomes paramount due to non-linear and inter-related problems of decreasing resources and the increasing complexities of warfare. For example, the US Navy can expect less manpower on fewer combatant ships (Office, 2006), increased amount of information, increased threat complexities, increased speed of warfare (U.S. Joint Chiefs of Staff 2000), increased speed of battle group response (Office, 2005), and single naval platforms having greater multiplexed missions across various war fighting commanders (U.S. Department of Defense, 2005b). This set of problems for the US Navy entering the 21st century is at the heart of systems thinking (Boardman and Sauser, 2008). There is an interwoven fabric of critical problem domains, and they all need to be addressed in a holistic force-wide systems' perspective. As a result, one possible resolution to the real-time complexities of a fast-paced dynamic threat environment requires a greater need for intelligent automation to complete missions and protect the fleet.

An ABMA is intended to provide ubiquitous sensing with automatic assignment of weapons against threats. This will decrease reaction time, increase accuracy, and optimize weapon utilization. In a naval context, the ABMA is different from other Battle Managers (BM) as the ABMA is concerned with composite threats at a force-wide level from the air, surface, and subsurface. It can assist decision commanders in real time with the assignment of combat systems and weapons doctrine; adjudication of Rules of Engagement; and integration, control, and deconfliction of multiple and disparate weapon responses to a threat (called Joint Fires). This also applies to the placement of ships and aircraft in response to a dynamic battle space environment for maximum mission effectiveness and to assure the safety of the battle force.

The importance of a composite BM is that the same friendly force asset, for example a cruiser, must prosecute all three-threat domains (Bailey, 1991) while present day BMs are concerned with one domain. It is very possible that in the course of prosecuting a threat in one domain, a threat will arise from another as an unintended consequence. For example, if a cruiser is prosecuting an air threat with its radar sensor and pairing to a missile, it could very well reveal its position to enemy sensors. The ABMA would consider unintended consequences given information about possible submarines in the area or other threat assets. A difficulty is the coordination and setting of priorities of an asset among different warfare commanders, albeit there is good communication among the control hierarchies to the composite warfare commander (Naval Reserve Intelligence Program, 1999).

At whatever level control is placed, the law of requisite variety states that the controller must be able to respond with at least as much complexity as the system it controls. This is an important facet. Historically, battle managers are centrally controlled. In the case of the ABMA, execution is decentralized composite coordination, in a cooperative manner among peers, with no component having control over others. For an ABMA, it is about logical centralized command and doctrine with decentralized control.

The ABMA executes using doctrines and Tactics, Techniques and Procedures (TTPs) which are principles and rules that the military uses to achieve national objectives tailored for local conditions. These rules provide guidance by networking people, organizations, systems, applications, data, and services in achieving shared awareness for a greater degree of self-synchronization behavior through emergence (Alberts, Garstka *et al.*, 2000; Alberts and Hayes, 2006) and as a result of horizontal fusion of systems. Horizontal fusion is a catalyst by combining and fusing data from traditional vertical stovepipe organizations and providing real time situational awareness throughout the battlespace. This results in battlespace awareness. Battlespace awareness is defined as "Knowledge and understanding of the operational area's environment, factors, and conditions, to include the status of friendly and adversary forces, neutrals and noncombatants, weather and terrain, that enables timely, relevant, comprehensive, and accurate assessments, in order to successfully apply combat power, protect the force, and/or complete the mission" (U.S. Department of Defense, 2001). The decentralized approach of the ABMA is a major transition. Since the 1990s battle management planning has been platform- centric through coordination and deconfliction on strong vertical organizational structures with limited horizontal connectivity between them. The ABMA provides a transition from hierarchical centralized preplanning that assumes a linear problem battlespace to dynamic decentralization that effects planning and execution of a nonlinear problem battlespace.

An ABMA requirement is to enable effect-based operations (EBO). EBO are sets of actions combining military and non-military methods, such as military, diplomatic, psychological, and economic instruments, to achieve a particular effect. The framework considers the full range of direct, indirect, and cascading effects directed at shaping the behavior of friends, neutrals, and foes by creating conditions to achieve objectives and policy goals (Davis, 2001) through self-synchronization. Self-synchronization is the interaction between two or more elements with a common rule set and shared awareness. A common rule set describes a common outcome of the interaction with a common intent or doctrine and shared awareness exchanges among the cooperating entities the information with regard to the dynamics of the environment via horizontal fusion. Self-synchronization allows for the execution of a mission, in the context of a changing environment, without the traditional hierarchies of Command and Control (C2). Self-synchronization is in accordance with a doctrine by low-level forces operating autonomously and retasking themselves through exploitation of shared awareness and doctrine. The doctrine in this case is known as Commander's Intent, and it is defined as "A concise expression of the purpose of the operation and the desired end state. It may also include the commander's assessment of the adversary commander's intent and an assessment of where and how much risk is acceptable during the operation" (U.S. Department of Defense, 2001). Self-synchronization is limited by traditional optimized systems. The need in this environment is for adaptive systems that provide for EBO which, in turn, provide the right information at the right time to the right decision-maker for the right weapon and the correct effect.

The overall objective of any BM is to manage battlespace awareness. This enables timely, relevant, comprehensive, and accurate assessments, so as to successfully apply combat power, protect the force, and provide the means by which commanders and force elements make better decisions faster. Making better decisions faster requires a thorough understanding of the environment in which they operate, relevant friendly force data, potential threats, and nonaligned elements that could aid in or subtract from overall objectives. ABMA is used in situational prediction as a projection of the current situation into the future to estimate enemy course of action (eCOA) and the effect of planned friendly forces or friendly COA (Young, 2004). It uses the plans developed but is designed to operate at a much faster tempo within a rapidly changing environment comprised of asymmetric threats, unexpected known threats, or conditions that may cause the loss of assets and risk the safety of the warfighter.

For a Naval ABMA, the major assumption is that most of its constituent systems are autonomous and heterogeneous. Battle Manager Systems may be comprised of several hierarchical BM systems that are local on a cruiser, regional BM on another cruiser or aircraft carrier, and a theater BM that may be resident on an aircraft carrier or land-based. The BM interfaces to organic sensors of various types ranging from different types of radars, to under water acoustics of various types, as well as sensors on other ships and aircraft. The BM obtains data of battlespace awareness from many sources that are local, regional, theater, or global. The BM will also manage weapon inventories, suggest weapons usage against a threat, and may manage their corresponding launchers to engage and launch against a determined threat.

The ABMA strategy is based on the autonomic compose ability of assets being sensors, weapons, and associated platforms. This shift from traditional platform-centric capabilities to network-centric capabilities assumes independent actors of the ABMA SoS in a continuous adapting system. This is critical as platform centric systems cannot leverage the ability to synergize. This is due to autonomous and coordinated interdependent systems. In the past, mission success was due to sequential and linear synchronized forces achieving a cumulative effect via attrition. An end state of mass effects is achieved without massing forces. For the ABMA SoS environment, mission success is accomplished through nonlinear, simultaneous, distributed, and parallel operations. The assets are self-synchronized to achieve synergistic effects based on the union of core competencies. The force units across echelons will no longer need the same degree of organic capabilities to achieve mission success because they rely on the interdependence of assets of other units.

Dynamic SoS ABMA Doctrine

An example of the importance of dynamic doctrine is the transformation of the characteristic of *connectivity* from platform-centric to network-centric and the capabilities derived from the collective whole via cooperation. As asynchronous threat possibilities increase and hostile counter measures become more sophisticated, the battle group must address individual ship capabilities and coordinate the sensors and weapons of individual ships into a cooperative and distributed Anti-Air Warfare (AAW), Anti-Submarine Warfare (ASW), and Anti-Surface Warfare (ASuW) system comprising a composite perspective.

A composite ABMA is required for a holistic and force-wide response to affect COA and counter eCOA. For AAW, self-synchronization is achieved by AAW officers talking to one another without the situation being revealed to an ASW officer who is responsible for locating and responding to subsurface threats. One cruiser will affect AAW operations while at the same time another cruiser will affect an ASW mission. These actions must be highly coordinated and must be agreed upon in cooperation. The characteristic of *belonging* plays heavily while the characteristic of *autonomy* is present. As a cruiser moves off to prosecute the ASW threat, radar coverage from a Naval platform may be available only if the ABMA understand the parameters of both missions and manages the missions in a constitutive SoS satisficing decision-making manner (Archibald, Hill et al, 2006). Neither mission is optimal, but the ABMA fulfills the necessary requirements affecting both missions.

The ships of the battle group operate to their own TTPs and general force-wide doctrines controlled by the lead command ship, a cruiser or an aircraft carrier, while each of the capital assets operates to a static doctrine. Changes in the doctrine are received via the ABMA based on new information and an updated force-wide doctrine is promulgated. Local TTPs adjust to accommodate the new force-wide doctrine. It is important to recognize that the characteristic of autonomy reinforces platform centric behavior but through the SoS characteristic of belonging, cooperation instills a network-centric behavior creating cooperative capabilities. This behavioral change introduces a paradox for autonomy, one of conformance and another of independence. The SoS characteristics are contrasted below and summarized in Table 1 between a Battle Group Ship platform-centric behavior and the ABMA distributed behavior.

The examination of the paradoxes (i.e. opposing forces) within each characteristic of ABMA heuristically indicates a point where those forces meet.

Autonomy

- **Conformance:** The ABMA creates a cooperative strategy to illicit capabilities that have their basis in the strength of the autonomy of the individual ships (Bailey, 1991)).
- **Independence:** Each ship within the battle group has its mission to prosecute a target with a priority of self-survival. A battle group may form defensive zones whereby assets in each zone perform their role (Bailey, 1991).

Belonging

- **Centralization:** Battle group assets are governed by their doctrines, deployment strategies, and TTPs. Their membership and missions are predetermined in defense of the battle group to fulfill commander's intent. They are central in nature with an emphasis of all capabilities integrated

Table 1. SoS Characteristics Contrast

SoS Characteristics	Battle Group Ship Platform	ABMA
Autonomy Independence Conformance	Managerial and operational independence; Predefined coordination capabilities.	Managerial and operational independence; Dynamic prediction and cooperation capabilities.
Belonging Decentralized Centralized	Centralized and decentralized. Directed membership and predetermined value of cooperation through TTPs and mission.	Decentralized with limited centralization of doctrine control.
Connectivity Network-Centric Platform-Centric	Platform-centric. Direct communications and information via predefined links and established TTPs and doctrine; Defined interoperability and interdependence. Integrated constituent systems.	Net-centric and interoperable constituents; Direct and indirect communications.
Diversity Heterogeneous Homogenous	Homogenous with other similar physical assets; Heterogeneous in independence in missions and capable of exhibiting heterogeneous behavior.	Independence; heterogeneous; Learning.
Emergence Indeterminable Foreseen	Foreseen. Indeterminable displayed through unintended consequences.	Learning; adaptable; indeterminable at composite force-wide level. Foreseen on platform level

onto a single platform based on how the Command and Decision system is designed versus derived capabilities through cooperation (Bailey, 1991).

- **Decentralization:** The battle group emphasizes the strength of its autonomous capabilities, but also extends its capabilities. The Air Intercept Controller (AIC) in a Combat Information Center (CIC) directs fighters from an aircraft carrier to engage air threats at great range extending the defensive zone of the battle group (Bailey, 1991) via constitutive SoS cooperative composeability.

Connectivity

- **Platform-Centric:** The composition of the systems to fulfill a mission is predefined by design-time requirements and is hierarchical in nature. The systems possess tightly fixed boundaries and behave as a single node within a larger system (U.S. Department of Defense, 2005). Platform-centric connectivity occurs via the process of engineering integration (DiMario, 2006).

- **Network-Centric:** The ABMA facilitates distributed groups of decision makers by providing the ability to cooperate in fulfillment of command intent in a collaborative environment. Connectivity occurs via interoperability of shared autonomous resources and affects a force-wide goal. The system structure is flat and distributed with dynamic boundaries. Communication couplings are loosely predefined (DiMario, 2006).

Diversity

- **Homogeneous:** There is little diversity among similar classes of battle group assets (Bailey, 1991). They have similar subsystems with variants as components evolve through their system's life cycles (O'Rourke, 2008). The ability to cooperate builds within the structure of the ship system via systems of communication and in the execution of rules.
- **Heterogeneous:** In Battle Group, assets are more diverse, and their capability to offer various services not germane to another platform is enhanced. Versatility to automatically decide between different types of sensors and weapons using the Observe-Orient-Decide-Act (OODA) loop methodology enhances the ABMA's ability to respond and adapt to changing threats (Caffall, 2005). Heterogeneity may not be different in the actual platform, but rather different in the capabilities they display as well as in the diversity of their missions.

Emergence

- **Foreseen:** Current systems are designed with detailed requirements and are evaluated operationally and technically. Emergence is defined as unwanted behavior whereby there is a deviation from requirements or a software bug (Caffall, 2005).
- **Indeterminable:** The ABMA SoS may elicit emergent unpredictable behavior based on its ability to decide on an indeterminate mix of sensors, weapons, and platforms in an indeterministic battle environment. Its primary goal is to achieve command intent, and the methods by which it achieves this may be indeterminable (Caffall, 2005).

Based on the analysis above of the five distinguishing characteristics and their paradoxes, Figure 6 demonstrates a qualitative depiction (i.e. snapshot) of the meeting point of the various forces within the dynamic

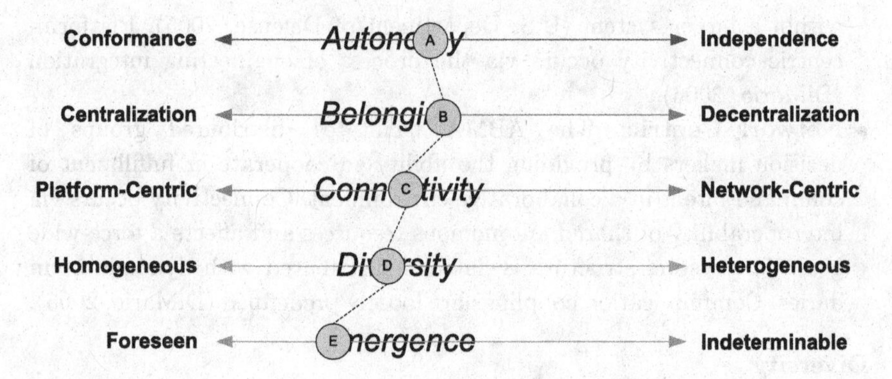

Fig. 6. SoS Characteristics Contrast.

doctrine of ABMA. The meeting point is dynamically shifting over time, as ABMA forces evolve. This qualitative representation is affirmed by a thorough review performed by five DoD contractor subject matter experts of BM modeling and simulation, fielded threat auto engagement systems, and fielded autonomous vehicles.

While it is difficult to indicate the exact meeting point on this scale using the qualitative analysis just described, the authors are currently investigating quantitative approaches to making such a determination. The Static SoS Doctrine shows us ways to implement corrective action to effectively influence each of the five distinguishing characteristics.

Static SoS ABMA Doctrine

The initial preset static doctrine of ABMA within a battle group is assessed and changed using the SoSE Management Conceptual Areas (i.e. modified FCAPS principles). These areas will be discussed in detail below and include risk management, configuration management, performance management, policy management, and resource management.

Risk Management: Risk management for the ABMA is an assessment process of SoS objectives and different types of risks that may potentially jeopardize the achievement of mission objectives. For example, it is absolutely necessary to collect and analyze information about unwanted emergent behavior, infrastructure, integration of multiple systems, and communications (Regal and Pacetti, 2008).

Configuration Management: Configuration management for the ABMA includes a process of identifying the current and likely mix of constituent systems within the battle group to accomplish a specific mission and to fulfill command intent. It is necessary to properly establish flexible provisions that will encompass not only the evolution of various capabilities but also the addition of similar capabilities as they emerge in the entire defense structure. The configuration management structure must find a meaningful method to correctly align asset sensors, weapons and platforms for sensor-weapons pairing in order to meet the needs of various command structures and missions (Donnelly and Galster, 2008).

Performance Management: Performance Management of the ABMA is based on the comparison processes of technical requirements of various battle group assets or SoS constituents and their actual implementation. These measurements include assessment of the external interfaces of the constituent systems; their compliance with the connectivity and communications protocols and standards; and their compatibility and cooperative ability with the other constituent systems in order to achieve command intent (Collens Jr and Krause, 2005).

Policy Management: Policy Management of any SoS is developed in response to requisite forms of management or execution or in response to unwanted behavior. There are various levels of policy management ranging from doctrine to TTPs, all of which must be congruently managed for the safety of the force and to fulfill command intent (Regal and Pacetti, 2008).

Resource Management: Resource Management uses a holistic perspective of SoS. It includes an examination of a specific system from different perspectives in the context of an overall SoS or, in this case, the battle group. In the ABMA case study, it is important to identify various objectives posted for each individual constituent system. This is so because they have their own autonomous mission, yet they may also participate in a cooperative manner that elicits cooperative capabilities (Donnelly and Galster, 2008). For example, an Unmanned Aerial Vehicle (UAV) plays an important role within the aviation system because it fulfills specific tasks of reconnaissance. At the same time, it is limited in its capacity to carry weapons. The aviation system is only part of the overall ABMA system. Therefore, depending on the particular context, it becomes necessary to establish a resource management strategy to achieve mission objectives.

"Satisficing" ABMA

Through examination of the ABMA case study, it becomes evident that the dynamic and static doctrines provide us with an opportunity to observe a newly developed management framework. The goal includes creating a framework that would allow us to effectively "satisfice" management demands of an ever-changing SoS environment. Figure 7 summarizes an important nexus between dynamic and static doctrines using SoSE Management Matrix (SoSEMM) (Gorod, Gove *et al.*, 2007) that would help us to achieve this significant objective.

Conclusion

A comprehensive review of the SoS related literature reveals a lack of effective methodology to achieve the objective of System of Systems Engineering (SoSE) (Gorod, Sauser *et al.*, 2008; DiMario, Cloutier *et al.*, 2008; Sousa-Poza, Kovacic *et al.*, 2008; Valerdi, Axelband *et al.*, 2008). The goal of SoSE includes successful engineering of multiple integrated complex systems.

In this paper, the authors proposed a SoS engineering management framework which will serve as a foundation for effectively "satisficing" SoS. The new method consists of the dynamic and static doctrines and introduces a new approach whereby requirements are met, but not optimized, allowing for the greater flexibility and adaptability of the system. This is described using dynamic and static doctrine feedback loops. In a dynamic loop, the system allows for the automatic adjustments of the preset conditions. The static loop provides us with an opportunity to intervene and make reactive changes in achieving "satisficing" goals. The static doctrine deals with the static states and their relationships while the dynamic doctrine manages the schema of change.

Further, we described the application of the proposed framework in the SoS domain using the case study of Auto Battle Management Aids (ABMA). By analyzing the case study, it is evident that the new framework provides a much needed holistic approach for effective ABMA management.

The proposed framework is the first crucial step towards structured SoS management. The research suggests promise in applying the "satisficing" method using the dynamic and static doctrine to the SoS domain. This will both reduce the complexity of management and increase operational effectiveness. In addition, the new framework also serves as a blueprint for the modeling and simulation of a SoS.

Static Doctrine

SoSEMM for Auto Battle Management Aids (ABMA) / Dynamic Doctrine	Risk Management	Configuration Management	Performance Management	Policy Management	Resource Management
Autonomy	Permits preservation of autonomous capabilities of ABMA constituent systems	Ensures protocols to effectively govern autonomous behavior of the ABMA constituent systems.	Monitors ABMA constituent systems mission to balance with commander's intent and assure own safety.	Establish the extent of cooperation of the ABMA constituent systems via doctrine modification, TTPs producing EBO fulfilling commander's intent.	Provides information which enables influencing the autonomy of the ABMA constituent systems to cooperate and fulfill commander's intent.
Belonging	Permits effective cooperation among ABMA constituent systems to meet commander's intent.	Provides guidance and coordination to unify cooperation of ABMA constituent systems to fulfill commander's intent.	Evaluates ABMA constituent system's contributions holistically to fulfill commander's intent.	ABMA manages constituent systems to fulfill commander's intent per TTPs.	Provides constituent systems of the ABMA of sensors, weapons, and capital assets in cooperative manner to fulfill commander's intent.
Connectivity	Allows and maintains interoperability of the ABMA constituent systems to fulfill commander's intent.	Establishes and maintains consistency in interoperability of ABMA assets and interfaces (e.g. protocols, standards, bandwidth, etc.)	Monitors and evaluates the constraints of syntactic and semantic interoperability of ABMA constituent systems (e.g. Link management).	Monitor and provide authorization and information assurance of the ABMA constituent systems.	Tracks and manages connected resources of ABMA. Capabilities provide varied levels of ship and force wide ability to access information and situation awareness as needed.
Diversity	Helps to evaluate diverse behavior of the ABMA constituent systems to provide holistic capabilities via cooperation.	Evaluate and direct logical and physical connectivity of disparate constituent systems and interfaces of the ABMA.	Monitors and evaluates the extent of disparate systems of fulfilling commander's intent with available ABMA constituent systems for holistic capabilities.	Restricts/permits via doctrine and TTPs behavior of ABMA constituent systems.	Provides information on ABMA diversified assets and systems including tactical position (e,g types of ships, weapons, weapon inventory, sensors, logistics tools, C41SR systems, etc.)
Emergence	Validates and verifies new behaviors based on monitoring, identifying, assessing, analyzing, and mitigating unintended consequences and reinforces holistic cooperation..	Shapes and bounds holistic behavior in order for ABMA to satisfice EBO to fulfill commander's intent.	Monitors and evaluates the extent of emergence that ABMA doctrine allows in accordance to commander's intent, as well as unintended consequences.	Bounds emergent behavior in order for ABMA to operate effectively, efficiently and safely.	Tracks new emergent properties both intended and unintended per COA and eCOA; provides performance trades of sensors and weapons per ABMA capital asset.

Fig. 7. Relationship between Dynamic and Static Doctrines of Auto Battle Management Aids.

References

1. Alberts, D. S., J. J. Garstka and F. P. Stein (2000) Network Centric Warfare: Developing and Leveraging Information Superiority, DoD Command and Control Research Program.
2. Alberts, D. S. and R. E. Hayes (2006) *Understanding Command And Control.* Washington D.C., DoD Command and Control Research Program (CCRP).
3. Aldridge, J., E. C. (2004) Report of the President's Commission on Implementation of United States Space Exploration Policy: A Journey to Inspire, Innovate, and Discover, NASA, US Government Printing Office.
4. Archibald, J. K., J. C. Hill, *et al.* (2006). "Satisficing Negotiations." IEEE Transactions on Systems, Man and Cybernetics, Part C: Applications and Reviews 36(1): 4–18.
5. Bailey, D. M. (1991) *Aegis Guided Missile Cruiser*, Motorbooks International.
6. Balci, O. (2008) Network-centric military system architecture assessment methodology, *Int. J. System of Systems Engineering*, **1**(1/2), 271–292.
7. Berryman, J. (2006) What defines 'enough' information? How policy workers make judgments and decisions during information seeking: preliminary results from an exploratory study, *Information Research*, 11(4).
8. Bertalanffy, L. v. (1969) *General System Theory: Foundations, Development, Applications.* New York, George Braziller, Inc.
9. Boardman, J. and B. Sauser (2006) System of Systems - the meaning of *of*, IEEE International Conference on System of Systems Engineering, Los Angeles, CA.
10. Boardman, J. and B. Sauser (2008). Systems Thinking: Coping with 21st Century Problems. Boca Raton, CRC Press.
11. Boulding, K. (1956) General Systems Theory: The Skeleton of Science, *Management Science*, **2**(3), 197–208.
12. Byron, M. (1998) Satisficing and Optimality, *Ethics*, 109 67–93.
13. Caffall, D. S. (2005) Developing Dependable Software For A System-Of-Systems, Monterey, Naval Postgraduate School. PhD.
14. Collens Jr, J. R. and B. Krause (2005) Theater Battle Management Core System Systems Engineering Case Study, Wright-Patterson AFB, Air Force Institute of Technology.
15. Davis, P. K. (2001) Effects-Based Operations: A Grand Challenge for the Analytical Community, Santa Monica, CA, RAND.
16. DeLaurentis, D. A., J.-H. Lewe and D. P. Schrage (2003) Abstraction And Modeling Hypothesis For Future Transportation Architectures, AIAA/CAS International Air and Space Symposium and Exposition: The Next 100 Years, Dayton, Ohio.
17. Dictionary (2001) Doctrine, *Webster's II New College Dictionary*, M. S. Berube, New York, Houghton Mifflin Company, 335.
18. DiMario, M. J. (2006) System of Systems Interoperability Types and Characteristics in Joint Command and Control, IEEE Conference on System of Systems Engineering, Los Angeles, CA.

19. DiMario, M. J., R. J. Cloutier, *et al.* (2008). "Applying Frameworks to Manage System of Systems Architecture." Engineering Management Journal, 20(4), 44–49 in press.
20. Donnelly, B. P. and S. M. Galster (2008) Prioritizations Taxonomy and Logic for Network-Centric Operations, *13th ICCRTS: C2 for Complex Endeavors*, Bellevue, WA.
21. Gorod, A., J. Gandhi, B. Sauser and J. Boardman (2008) Flexibility of System of Systems *Global Journal of Flexible Systems Management*, in press.
22. Gorod, A., R. Gove, B. Sauser and J. Boardman (2007) System of Systems Management: A Network Management Approach, IEEE International Conference on System of Systems Engineering, San Antonio, TX.
23. Gorod, A., B. Sauser and J. Boardman (2008) System of Systems Engineering Management: A Review of Modern History and a Path Forward, *IEEE Systems Journal*, **2**(4):1–16.
24. Gorod, A., B. Sauser and J. Boardman (2008b) Paradox: Holarchical View of System of Systems Engineering Management, IEEE International Conference on System of Systems Engineering, Monterey, CA.
25. Keating, C., R. Rogers, R. Unal, D. Dryer, A. Sousa-Poza, R. Safford, W. Peterson and G.Rabaldi (2003) Systems of Systems Engineering, *Engineering Management Journal*, 15(3), 36–45.
26. Kiyota, T., Y. Tsuji and E. Kondo (2003) Unsatisfying Functions and Multiobjective Fuzzy Satisficing Design Using Genetic Algorithms, *IEEE Transactions On Systems, Man, And Cybernetics - Part B*, 33(6), 889–897.
27. Krygiel, A. J. (1999) Behind the Wizard's Curtain: An Integration Environment for a System of Systems, *CCRP - C4ISR Cooperative Research Program*, Vienna, VA.
28. Naval Reserve Intelligence Program. (1999) MODULE 3—BATTLEGROUP COMMANDERS & THE CWC CONCEPT, from http://fas.org/irp/doddir/navy/rfs/part03.htm.
29. Office, U. S. G. A. (2005). Military Readiness: Navy's Fleet Response Plan Would Benefit from a Comprehensive Management Approach and Rigorous Testing. Washington D.C., U.S. Government Accountability Office.
30. Office, U. S. G. A. (2006). Defense Acquisitions: Challenges Associated with the Navy's Long-Range Shipbuilding Plan. Washington D.C., U.S. Government Accountability Office.
31. O'Rourke, R. (2008) Navy DDG-1000 and DDG-51 Destroyer Programs: Background, Oversight Issues, and Options for Congress, C. R. Service, Washington D.C., Congressional Research Service.
32. Pettit, P. (1984) Satisficing Consequentialsm, *Proceedings of the Aristolean Society*, 58 165–176.
33. Regal, R. and D. Pacetti (2008) Collaborative Technologies for Network-Centric Operations, Concepts, Theory and Policy, C2 Architectures, *13th ICCRTS: C2 for Complex Endeavors*, Bellevue, WA.

34. Reid, P. P., W. D. Compton, J. H. Grossman and G. Fanjiang (2005) *Building A Better Delivery System: A New Engineering/Health Care Partnership.* Washington D.C., The National Academies Press.
35. Saunders, T., C. Croom, W. Austin, J. Brock, N. Crawford, M. Endsley, E. Glasgow, D. Hastings, A. Levis, R. Murray, D. Rhodes, a. Sambur, H. Shyu, P. Soucy and D. Whelan (2005) System-of-Systems Engineering for Air Force Capability Development: Executive Summary and Annotated Brief.
36. Sauser, B. and J. Boardman (2006) From Prescience to Emergence: Taking Hold of System of Systems Management *27th Annual National Conference of American Society for Engineering Management*, Huntsville, AL.
37. Sauser, B., J. Boardman and A. Gorod (2008) SoS Management, *System of System Engineering - Innovations for the 21st Century*, M. Jamshidi, Hoboken, NJ, Wiley & Sons.
38. Schmidtz, D. (1996) *Rational Choice and Moral Agency*, Princeton University Press.
39. Simon, H. (1955) A Behavioral Model of Rational Choice, *Quarterly Journal of Economics*, 69(1), 99–118.
40. Simon, H. (1956) Rational Choice and the Structure of the Environment, *The Psychology Review*, 63 129–138.
41. Sousa-Poza, A., S. Kovacic and C. Keating (2008) System of systems engineering: an emerging multidiscipline, *Int. J. System of Systems Engineering*, **1**(1/2), 1–17.
42. U.S. Congress (1988)*SDI: Technology, Survivability, and Software.* Washington DC, US Government Printing Office.
43. U.S. Department of Defense (2001) Joint Pub 1-02, *Department of Defense Dictionary of Military and Associated Terms*
44. U.S. Department of Defense (2005) FORCEnet: A Functional Concept for the 21st Century, Washington DC.
45. U.S. Department of Defense (2005b). Command and Control Joint Integrating Concept. Washington D.C.
46. U.S. Joint Chiefs of Staff (2000). Joint Vision 2020, Government Printing Office, Washington D.C.
47. U.S. Joint Chiefs of Staff (2005) Joint Capabilities Integration and Development System: CJCSI 3170.01E.
48. Valerdi, R., E. Axelband, T. Baehren, B. Boehm, D. Dorenbos, S. Jackson, A. Madni, G. Nadler, P. Robitaille and S. Settles (2008) A research agenda for systems of systems architecting, *Int. J. System of Systems Engineering*, 1(1/2), 171–188.
49. Young, B. W. (2004) A C2 Systems for Future Aerospace Warfare, *2004 International Command and Control Research and Technology Symposium (CCRTS): The Power of Information Age Concepts and Technologies*, San Diego, CA.
50. Zsambok, C. E. and G. Klein (1997) *Naturalistic decision making*, Lawrence Erlbaum Associates.

Appendix G

SYSTEM OF SYSTEMS COLLABORATIVE FORMATION — SUBMITTED FOR PUBLICATION

MICHAEL J. DIMARIO, *Member, IEEE*, JOHN T. BOARDMAN,
BRIAN J. SAUSER, *Member, IEEE*

The formation of a system of systems from a collection of interdependent systems does so in response to a collaborative mechanism. The nature of the constituent systems that compose the system of systems is a paradox by which they maintain an autonomous behavior and yet join the collective in collaboration. A social function can describe the system of systems mechanism that considers the value of collaborative relationships, and stresses preferences for action rather than an independent action of the individual systems. The constituent systems of the system of systems balance the risk of belonging with that of autonomy. This decision dichotomy is an example of typical decision makers that must balance cost and risk against achieving goals in a satisficing environment. A process beginning with a global goal serves to construct a Multicriteria Decision process creating the set of alternatives representing choices forming a collective social function that binds the system of systems. A case study demonstrates a simple application of independent autonomous systems collaborating to affect a holistic response with positive results.

Keywords: System of Systems, SoS, SoS Characteristics, Satisficing, Analytical Hierarchy Process, AHP, Multiple Criteria Decision Analysis, MCDA.

Introduction

The engineering and control of emergent capabilities of large complex systems comprised of numerous systems is an apparent unattainable goal of technologists and social infrastructure planners. This can be similarly described as a metaphor of an "invisible hand" from work of Adam Smith in Book IV of the *Wealth of Nations* [2]. The "invisible hand" metaphor is intended to explain that actions have unintended consequences, are not controlled by a central command authority, and have an observable and patterned effect on the processes and systems. In engineering of systems, we understand and explain their cooperative nature via their architecture, design, and control as characterized by a set of interconnections through a common goal. Yet in real-time, this cooperative control is not well understood let alone intentionally designed [3, 4].

We now read that many systems are being described as a system of systems (SoS) that comprise numerous constituent interdependent systems and have been described as an integration of complex systems [5]. This description supposes that the systems are autonomous and heterogeneous forming partnerships whereby their interoperability relationship produces capabilities or unintended consequences because of emergent behavior. The emergent behavior does not originate from any single individual constituent system nor deduced by properties of the collective constituent systems. The interoperability relationships of the constituent systems create new behaviors of the holistic system.

A review of General Systems Theory (GST) reminds us of the holistic concept of systems as a science of wholeness whereby there are general systems laws, which apply to any system of a certain type independent of its properties, classified as *summative* and *constitutive*. The summative system owes its capabilities to the summation of the characteristics and sub-capabilities of its elements. The properties of the elements are universally the same in all environments and have no relationships. The constitutive system has capabilities that are greater than the summation of its elements because the effects are dependent upon the context of their properties and their relationships [6]. Boulding described this concept as a SoS which may perform the function of a *gestalt*, which is a pattern so unified as a whole that its properties cannot be derived from its parts [7].

An important idea to be made with respect to the spectrum of the types of systems is that a SoS is a system, but not all systems are a SoS [8]. A SoS, unlike traditional or ordinary single systems, is proposed to have unique attributes and are discussed extensively in the literature [9–11]. A commonly discussed set of characteristics that are used here are autonomy, belonging, connectivity, diversity, and emergence as they contrast a system and a SoS [8]. A SoS is comprised of autonomous constituent systems fulfilling an objective as independent systems but are interdependent via interoperability to fulfill holistic objectives. The nature of the constituent systems of the SoS in response to a mechanism of collaboration enters a paradox by which it maintains an autonomous behavior and joins the collective in collaboration. SoS emergent behavior and capabilities are a product of the interactions that are greater than the sum of the independent actions of the constituent elements describing a constitutive SoS. Behavior arises from the organization of the elements, vis-à-vis the constituent systems without having to identify the properties of the individual elements. For example, the U.S. National Transportation System is a SoS comprised of an organization of large autonomous disparate transportation systems such as airlines, trains, and trucks [12].

This paper discusses a constitutive SoS framework in regard to new behaviors that emerge as a result of a mechanism of collaboration that is architected to affect a holistic capability among autonomous systems. The management and design of SoS architectures are of interest as individual systems become ever more interoperable with other systems [13]. The ability to design and control a SoS architecture to elicit capabilities not germane to any one individual system becomes paramount [14]. This paper builds upon the following two constructs: first, the theory of satisficing whereby systems reduce their autonomy and optimization to levels of acceptable risk to collaborate and in so doing describes the nature of SoS constitutive mechanisms. This establishes a mechanism framework whereby the paradox of autonomous and yet cooperative systems is described, and thus architected, to produce SoS capabilities. Second, the grouping of autonomous systems that choose to collaborate in a dynamically forming SoS is a result of changes in the autonomous systems' environment or their own utility using Multicriteria Decision Aids (MCDA) as a proxy collaborative system's decision process.

As a result, we seek to provide two unique contributions to the body of knowledge of SoS theory, management, and engineering. First, assesses the dynamics of a SoS through the use of five core characteristics, namely autonomy, belonging, connectivity, diversity and emergence; and second, describes a mechanism of collaboration whereby the characteristics of autonomy and belonging are satisficing for the SoS constituents and the resultant emergent behavior provides value for the observer. We actualize our approach with a case study that reveals that the performance of autonomous collaboration as a SoS exhibits greater capability than of the autonomous units performing independently. We conclude with a discussion of the implications of this research and potential future research directions.

Constitutive Satisficing SoS Mechanism

SoS is realized through the weaving of constituent systems via multiple connections. The connections are direct as logical or physical paths and indirect based on effects of influence such as how politicians are motivated by their constituency [15]. Systems are bounded as they rely on a hierarchical structure, central control, central data, and a high degree of trust. These properties are typically that of an independent and stable system optimized for improved performance. However, there is a paradox for optimized systems in a SoS context whereby too much efficiency impairs the ability to respond as the constituent systems lose their dynamic or adaptive capabilities. Optimized systems are inflexible, as their interfaces to other systems have fixed boundaries and tight coupling represented by a summative system versus dynamic boundaries and loose coupling represented by a constitutive system. Cooperative mechanisms of a constitutive SoS are likely to be satisficing, defined as a solution with acceptable intrinsic cost trade-off [16].

The satisficing process, first suggested by Herbert Simon [17], of a holistic system includes the management of the constituent systems to achieve a favorable result for the user [18]. The process seeks an adequate versus an optimized solution because it is not possible to optimize for an individual system and the whole system simultaneously. It is difficult to define what is best for the group in terms of individual preferences [19]. Satisficing allows for degrees of fulfillment, whereas optimization is absolute.

We theorize that mechanism design may give us further insights into satisficing for SoS. Mechanism design found in economics, political science, and game theory is the subject of designing rules of a system to achieve a specific outcome, even when the system may be autonomous and of self-interest [20]. For game theory, the purpose of mechanism design is to architect the rules of the environment or the rules the agents are to play for maximization of the collective good of the agents. A mechanism that allows for a constitutive SoS provides a holistic description and a basis of decision to how and why systems as a collective choose to cooperate and define the nature of the SoS characteristics. A fundamental premise adopted from game theory is that not all constituent systems behave the same and are autonomous and heterogeneous. The environment by which systems cooperate, as well as the systems themselves, is designed and managed to achieve a common goal even though the holistic behavior appears to be amorphous [18].

The constitutive SoS collaborative mechanism may best be described as a global utility function (U). The SoS global utility function exists via collective interdependent systems' architecture of composeability by methods of self-organization [14] and drawn from utility theory [21]. Rational systems weigh the advantages and disadvantages of cooperation and collaboration based on known and predicted facts that are subject to change. The SoS utility is a function of the value delivered to the users. To reach this holistic value, the individual participating systems weigh their own self-interest gains against the costs of cooperation. As long as the acceptable benefits are greater than the costs, constituent systems will cooperate in a satisficing manner [22, 23].

The maximization of a system's self-interest or utility function u_i underlies all economic theory and applications. Combined with mechanism design, there emerges a framework by which SoS comprised of multiple disparate systems may be analyzed. It becomes the looking glass to reveal what is desirable and what is possible in the organization of systems [24]. The natures of the interdependent constituent systems of a SoS by which they interoperate define the mechanism in the context of the acts of coordination, cooperation, and collaboration and summarized per the SoS characteristics in Table 1.

Table 1. SoS Interoperability Spectrum.

	Optimizing		Satisficing
SoS	**SoS Summative**	$\Longleftrightarrow$	**SoS Constitutive**
SoS Characteristic	**Coordination**	**Cooperation**	**Collaboration**
Autonomy	Directed	Negotiated with self-interest	Negotiated relinquishing self-interest
Belonging	Directed; optimized	Negotiated with self-interest	Negotiated relinquishing self-interest; Satisficing
Connectivity	Directed and bounded	Negotiated	Negotiated
Diversity	Planned with tightly bounded interfaces	Planned; Negotiated interfaces	Robust
Emergence	Foreseen	Foreseen with unintended consequences	Adaptable; Unforeseen

System of Systems Decision-Theoretic Approach

A fundamental basis of decision theory is a system that is rational if and only if it chooses actions that yield the highest utility. A system's belief state is a representation of the probabilities of all possible actual states of the environment. The choice of an action of a system state is determined by the comparison of utilities when there is risk or uncertainty. For rational systems, the choice of an action is with the maximum expected utility (*EU*). The holistic collective benefit through the individual system's participation and cooperation is reached when each system adopts a solution that maximizes its own utility function (u_i) or self-interest. The decision to maximize the system's EU is in response to its perception of the environment. If probabilities are unknown, a number of methods have been proposed such as choose any systems of weights or maximize the minimum results [25].

The maximization of the individual constituent system's utility function (u_i) is shown through the SoS characteristic of *autonomy* whereby each system does not know about the whole problem being solved, just that its own utility is maximized. The constituent system maximizes its autonomy in fulfillment of its goals and participates in the collective in self-interest resulting in a global utility function. However, the simple summation of the individual constituent system's utility function does not necessarily contribute to the utility function of the whole. It results in a summative SoS and the constituents degrade the global utility because actions of each system effects the other. However, either an individual system may maximize their own EU or that of the SoS, but not both; a situation of optimizing versus satisficing. If the SoS is to optimize its EU then the solution may not be acceptable to the individual constituent systems. However, a dominant strategy may exist whereby a strategy of cooperation yields a higher utility no matter what the other constituent systems do.

An SoS mechanism is achieved when a social function provides choices for the constituent systems to fulfill their EU in a holistic interaction. The social function in a SoS mechanism articulates the utility function that describes the holistic SoS environment concept represented in Figure 1. The collaborative SoS mechanism is created by establishing a structure by which the constituent systems have an incentive to behave according to a social function concept. In cooperative systems, the constituent systems are allowed to make binding agreements. A SoS mechanism implementation creates the social function executing the allowable policies that imparts to

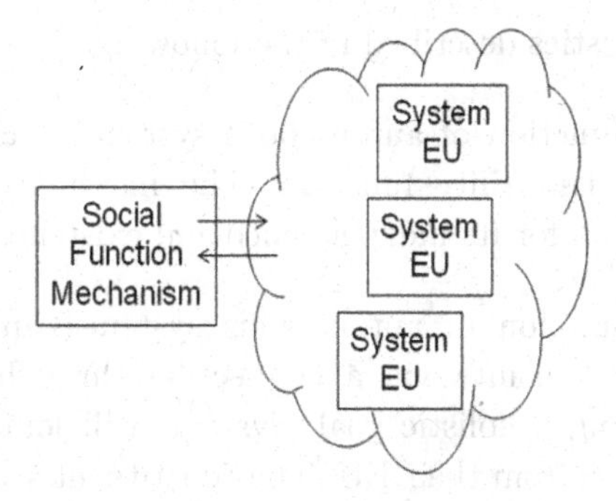

Fig. 1. SoS Social Function Mechanism.

utility theory the reasons for systems to satisfy their EU. The mechanism may be as simple as a voting scheme whereby systems vote if certain rules of self-interest are satisfied and the winner is determined against another set of rules establishing the social function.

Influence of System of Systems Characteristics in Mechanism Formation

The characteristics of SoS play important roles in the SoS mechanism design and social function. There is an implicit feedback structure shown in Figure 2 whereby characteristics influence the existence and behavior of

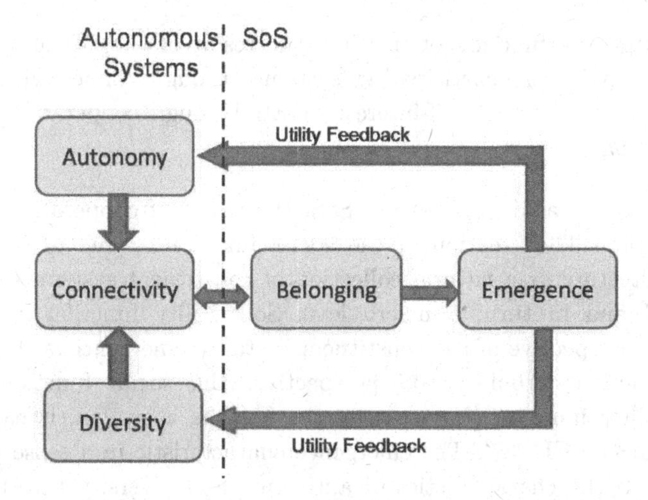

Fig. 2. Influence of SoS Characteristics in Social Function Mechanism.

the other characteristics described in the following:

Autonomy — a function of autonomous systems, a characteristic must exist to maximize its utility function. The maximization of its utility function is the reason for its independence and existence.

Diversity — a function of autonomous systems, an assumption that a SoS maximizes its utility via a connection through cooperation and collaboration forming a holistic goal. Systems will join with another out of self-interest driven from their EU. The constituent system is objectively different from other systems and is autonomous in their exhibition of their heterogeneity. The SoS mechanism responds by the gathering of unlike systems fulfilling the intended social function [26].

Connectivity — a function of autonomous systems, the means of communication and interoperability between the constituents of the SoS to enable a social function. The means may be any function that causes another to behave differently ranging from indirect influence to direct transmission of data. The connectivity within a SoS is described by [27] as "not presciently designed but emerges as a property of present interactions among holons." The connection to be made must also be available via the mechanism and the SoS characteristic of belonging. The characteristics of connectivity and belonging are tightly intertwined as a connection would not be made if it did not belong [3, 26].

Belonging — a function of the SoS, the result of the EU of autonomy and diversity of the participating systems through connectivity finding a reason out of mutual self-interest of EU through cooperation forming collaboration [28].

Emergence — a function of the SoS, the result of cooperation forming collaboration. The creation of the social function is due to intentional SoS architecture or a natural collection of constituent systems exercising their EU and in turn, a macro level SoS utility function is created. From the perspective of the constituent systems, emergence is the holistic social function. From a SoS perspective, the social function is the global utility function (U) whereby the SoS, as a system, is capable of maximizing its EU [27]. The emergence characteristic in a sense provides feedback to the characteristics of autonomy and diversity throttling the degree of autonomy and heterogeneity to elicit collaboration to minimize unintended consequences and provide emergent capabilities valued by the observer.

Satisficing System of Systems Mechanism Design Social Function

Social functions are not a logical consequence of rational systems based on self-interest. Group behavior is not optimized when individual constituents are optimized in pursuit of maximizing their EU as is done in conventional von Neumann-Morgenstern game theory [21]. Satisficing game theory permits the direct consolidation of group interests, and as such SoS goals, as well as individual interests. Satisficing replaces individual interests for group interest via a holistic model of inter-system relationships [29] creating collaboration. The SoS characteristics become evident in a group of systems with common goals in a satisficing manner of fulfilling their self-interest EU and the group's social function EU referred by Stirling as a private utility and social utility respectively [30].

The group's social function that elicits the common preferences of the participating systems creating interdependence of collaboration comprises the totality of preference relationships and the sense of belonging. The system's preferences are conditional in meeting requirements of collaboration. System interdependence is established as linkages are established with coherent system EU preferences resulting in SoS capabilities. The mechanisms by which preferences are selected comprise the social function. The interdependence function, a sub function of the social function, comprised of the SoS characteristics of belonging, connectivity, and diversity are based on the participant's ability to select or reject preferences to satisfy its EU. An example is the U.S. health care system whereby the system does not exist if it were not for the various systems that comprise it such as pharmaceutical companies, hospitals, insurance companies, physician offices, medical training organizations, and emergency medical response teams. The disparate organizations combine in a unique and adaptive way by which they all influence one another at various levels to be functional in their autonomy and diversity and interdependent in their emergent collaboration via connectivity and belonging. The participating systems of the health care system have released a portion of their autonomy and thus their optimization to collaborate. The decisions in choosing the available options for the participating health care elements define their satisficing set of preferences [31].

Multicriteria Decision Making

The satisficing system decision-making capacity is optimistic in selecting preferences in meeting its goals regardless of risk, cost, and resources or is pessimistic and rejects preferences in achieving cooperative goals and conserves resources, reduces cost, and lowers risk. The MCDA process generalizes a prioritization approach by evaluating a finite number of actions versus criteria. An optimal solution to a MCDA problem is one that maximizes each of the objective functions simultaneously. It is not likely to obtain an optimal solution that satisfies all criteria due to problems of conflicting objectives and preferences. The outcome of applying preferences represents a compromise solution that reconciles all the criteria and is satisficing. A satisficing solution is a set of preferences, which meets all the aspiration levels of each of the attributes. At a minimum, the final solution is reached by screening out the unacceptable solutions via lower priority preferences. It provides an objective analysis of alternatives against a goal and the supporting choices of criteria, alternatives, and relative weights via preferential trade-offs [32].

An important task for decision-making includes the evaluation of alternatives and subsequent preferences. The Analytical Hierarchy Processes (AHP), developed by Thomas Saaty [33], is used to derive weights in multicriteria decisions such as in economics, construction, and decisions regarding environmental effects of human activity [32]. The AHP process is used in this investigation as a simple proxy process to execute a SoS social function. It is a method to understand a complex system from a holistic perspective. It does so through deductive systems analysis by examining the functions of the components versus the components themselves. It is an indirect weight elicitation technique using paired comparisons regarded as the most common and superior technique to measure decision choices across multiple choices [34]. It considers the problem or goal holistically including constituent systems' functions within their environment as a network and within a hierarchy.

The SoS collaborative approach shown in Figure 3 portrays systems that negotiate their utility (u) to achieve a global utility (U) or social function. The systems thus change their state and enter a collaborative union demonstrated here using MCDA in pairwise comparisons of preferences. Their utility values are modified thus their states change. They may re-enter into a collaborative state with a new utility function to create a new social function. This becomes a dynamic iterative process whereby autonomous system participation is dependent on satisficing their utility

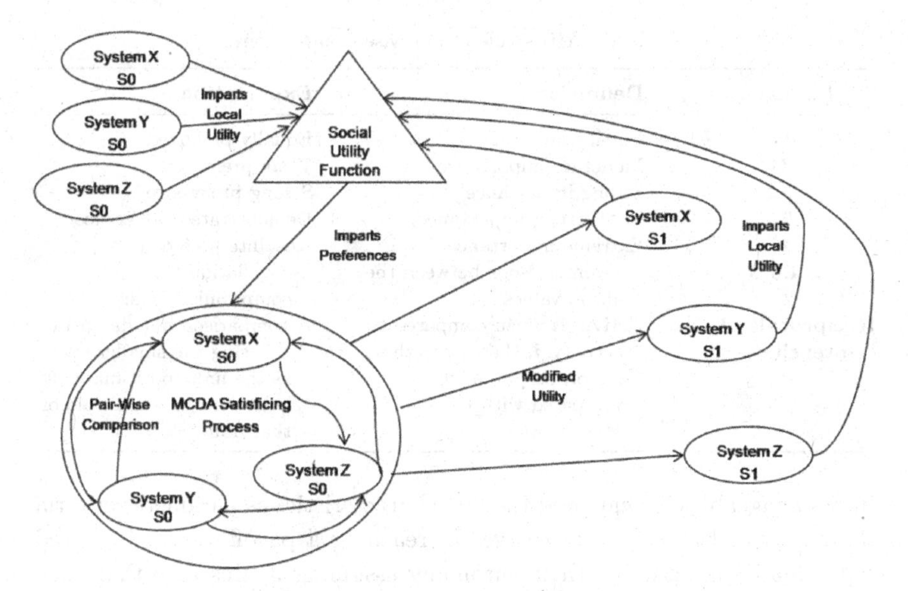

Fig. 3. System State Transition Concept.

function as well as satisficing a social function of the SoS. Thus, a social function is created out of autonomous self-interest [34].

AHP decomposes complex problems into several hierarchies with the last hierarchy comprised of the preferences to accomplish the goal. At each level of the hierarchy, the mechanisms of decision assign relative weights via a pairwise comparison. AHP is based on three principles of decomposition, preference measurement, and analysis. For the SoS environment shown in Figure 4, decomposition provides a breakdown of the social function

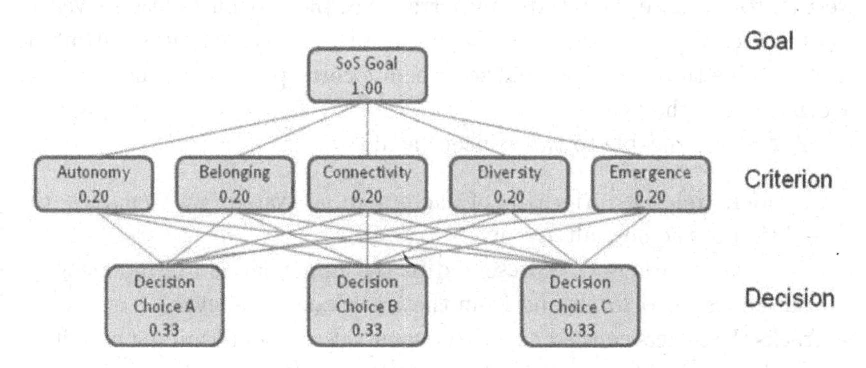

Fig. 4. SoS Characteristics Ideal Criterion AHP Structure.

Table 2. AHP Scale of Pairwise Comparisons.

Intensity	Definition	Explanation
1	Equal importance	Equally preferred
3	Moderate importance	Weak preference
5	Strong importance	Strong preference
7	Very strong importance	Demonstrated preference
9	Extreme importance	Absolute preference
2,4,6,8	For comparisons between the above values	Interpolation of a compromised judgment
Reciprocals of intensities	Activity i when compared to activity j, then j has the reciprocal value when compared with i	A comparison mandated by choosing the smaller unit as the unit to estimate the larger one as a multiple of that unit

into manageable elements and the SoS characteristics as parameter criteria decomposing to choice alternatives in reaching a payoff with all treated individually maintaining their autonomy assuming in this case that each have equal contributions.

AHP preferences measured on a semantic nine-point scale for pairwise comparisons shown in Table 2. The comparisons are entered into a pairwise comparison matrix with respect to their relative importance to their control criterion. A relative ratio scale of measurement is derived from a pairwise comparison of the elements at each level in the hierarchy based on the influence of the element in the above level for AHP or in the connecting node for ANP. Local priority vectors derived for all comparisons and the local priorities are synthesized to derive the global priority used in making the final decision of a preference. The weights reflect the priority of the element. An Eigen value is determined to find the normalized vector of weights to promote the relative importance of the attributes at all levels of the hierarchy. The normalized weights of all the hierarchy levels combine to determine the unique normalized weights corresponding to the last level accomplishing the goal.

AHP is applicable to this concept as it:

- Considers relative priorities of factors in a system and enables the selection of the best alternative,
- It assumes the factors represented in the upper levels are independent from all its lower levels and from the criteria at each level,
- Tracks the logical consistency of judgments used in determining priorities,

- Provides a scale for measuring intangibles and a method for establishing priorities, and
- Assumes interdependence of elements in the system and does not require linearity.

Case Study of an Auto Battle Management Aid System of Systems

A case study of a concept of operations of an Auto Battle Management Aid (ABMA) reveals the complexities of a dynamic SoS and the real-time management and doctrine of such a system [17]. Battle managers were recognized early in the development of the Strategic Defense Initiative Ballistic Missile Defense (SDI BMD) system of the 1980s to reduce time, cognitive load, and to manage autonomous computer systems [36]. Battle management is defined as "The management of activities within the operational environment based on the commands, direction, and guidance given by appropriate authority" [37]. The cognitive load is associated with the networking of systems logic, information processing, information exchange, and services as well as the semantic interpretation and translation of the exchanged data among the systems. Autonomic computing is required to reduce the costs of increased complexity and to respond to a large and asymmetric threat-enriched environment quickly. It also mitigates the need for operational interaction by automating the response of the network event or actions based on a centralized doctrine. The autonomic response recognizes different classes of events and their thresholds for appropriate response. The automaticity is based on the tempo of various levels required for a response. For example, for the U.S. Navy the tempo may range from months to weeks at command centers based on land and days to hours to seconds within an at sea Navy battle group.

The objective of an ABMA and value is to reduce cognitive load by reducing the decision time required to respond in a rapidly evolving dynamic threat environment. An ABMA becomes paramount due to non-linear and inter-related problems of decreasing resources and the increasing complexities of 21st century warfare [38]. For example, the U.S. Navy can expect less manpower on fewer combatant ships [39], increased amount of information, increased threat complexities, increased speed of warfare [38], increased speed of battle group response [40], and multiplexed missions of single naval platforms across various war fighting commanders [41]. This set of problems for the U.S. Navy entering the 21st century can only be addressed through holistic systems thinking [25] because of the nature of a fabric of critical problem domains to be solved in a holistic force-wide systems' perspective. As a result, one possible resolution to the real-time complexities of a fast-paced dynamic and indeterministic threat environment requires a greater need for intelligent automation to complete missions and protect the fleet.

An ABMA intends to provide ubiquitous sensing with automatic assignment of weapons against threats. This decreases reaction time, increases accuracy, and optimizes weapon utilization. In a naval context, the ABMA is different from other Battle Managers (BM) as the ABMA is concerned with composite threats in a force-wide context from near-space, air, surface, and subsurface. The ABMA provides deconfliction for multimission Navy platforms as they execute antisubmarine warfare (ASW), antisurface warfare (ASuW), antiair warfare (AAW), and ballistic missile defense (BMD). It can assist decision commanders in real-time with the assignment of combat systems and weapons doctrine; adjudication of Rules of Engagement; and integration, control, and deconfliction of multiple and disparate weapon responses to a threat. The ABMA planning and real-time response includes the placement of ships and aircraft in response to a dynamic battle space environment. This assures a goal of maximum mission effectiveness and safety of the battle force.

The principle of the ABMA is a shift from a traditional system platform centric behavior and response to a force-wide centric response summarized in Tables 3 and 4. The Conceptagon in Table 3 is used to distill a system versus a SoS as a set of seven triples representing the systems engineering lexicon [25] and Table 4 is represented by the SoS characteristics. Platform centricity and thus autonomy is applicable if the battle group asset is "alone and unafraid." A platform centric behavior is not applicable if the asset is "afraid and not alone." It is threatened and at high risk of damage and loss of life and must seek assistance from other battle group assets. Platform centric assets are truly autonomous with little ability to relinquish their autonomy in a collaborative manner. They require extensive coordination for any kind of cooperation.

Force-wide SoS capabilities are of a run-time nature albeit their interfaces are open to disparate connectivity unless they are planned for a force-wide response to a force-wide threat. For example, if a cruiser is prosecuting an AAW threat with its radar sensor and pairing to a missile, it could very well reveal its position to other hostile sensors. The ABMA would consider unintended consequences given information about

Table 3. Conceptagon of Platform vs. Force-wide Centricity.

Conceptagon	System Platform Centricity via Optimizing	SoS Force-wide Centricity via Satisficing
Function Structure Process	Composing non-autonomous constituents or predefined autonomous systems; Hierarchical	Composing autonomous constituents; Flat; Distributed
Relationships Wholes Parts	Predefined	Dynamically defined; Composeable
Emergence Hierarchy Openness	Predicted	Unpredicted
Communication Command Control	Central	Distributed
Boundary Interior Exterior	Fixed	Dynamic
Transformations Inputs Outputs	Tight; Predefined	Dynamic; Undefined
Harmony Variety Parsimony	System utility function	Social utility function

possible submarines in the area or other threats. A logical difficulty is the coordination and setting of priorities of an asset among different warfare commanders. At whatever level control is placed, the law of requisite variety states that the controller, in this case a warfare commander, must be able to respond with at least as much complexity as the systems they control. This is an important facet. Historically, BM is centrally controlled. In the case of the ABMA, execution is decentralized composite coordination, in a collaborative manner among peers and summarized in Table 4.

Mechanism Case Study Scenario Simulation

A scenario based SoS mechanism ABMA model is created to demonstrate the interactions of alternatives using the SoS characteristic of autonomy as well as other pseudo criteria. In this scenario, SoS capabilities emerge via collaboration producing a force-wide response. In order to drive emergence,

Table 4. Contrast of SoS Characteristics of Control vs. Forms of Decentralized Control.

SoS Characteristics	Optimize Combatant Platform Central Control	Satisfice ABMA Decentralized Control
Autonomy	Managerial and operational independence; Predefined coordination capabilities.	Managerial and operational independence.
Belonging	Centralized and decentralized. Directed membership and predetermined value of cooperation through TTPs and mission. Predefined interdependence.	Decentralized. Collaboration due to shared doctrine and missions. Predefined and composeable interdependence.
Connectivity	Platform-centric. Direct communications and information via predefined links and established TTPs and doctrine; Defined interoperability and interdependence. Integrated constituents.	Netcentric and interoperable constituents; Direct and indirect self-organizing communications by allowed links. Indirect communications by effects or stigmergy.
Diversity	Homogenous with other similar physical assets; Heterogeneous in independence in missions and capable of exhibiting heterogeneous behavior.	Independence; heterogeneous; Evolving intelligence. Robust opportunities for emergent capabilities.
Emergence	Foreseen. Displayed through unintended consequences.	Evolving intelligence; adaptable; indeterminable.

the characteristic of belonging is required because assets must impart their utility function and relinquish levels of autonomy. Modulating the characteristic of belonging is performed by modulating the characteristic of autonomy whereby autonomy is determined in pair-wise comparisons of other preferences shown in Figure 5 based on a snapshot of a real-time iterative AHP process shown in Table 5. As shown in Table 5, the criterion of a re-engagement is of priority followed distantly by inventory and radar resources. To be able to reengage a threat means that an earlier attempt to neutralize a threat had failed and having the resources and capability for a second attempt becomes the highest priority. The approach provided in the ABMA case study is a simple AHP decision process for a SoS that articulates a social function based on preferences that fulfills the utility function of the warfare commanders and the autonomous systems, in this case as neutralizing air threats.

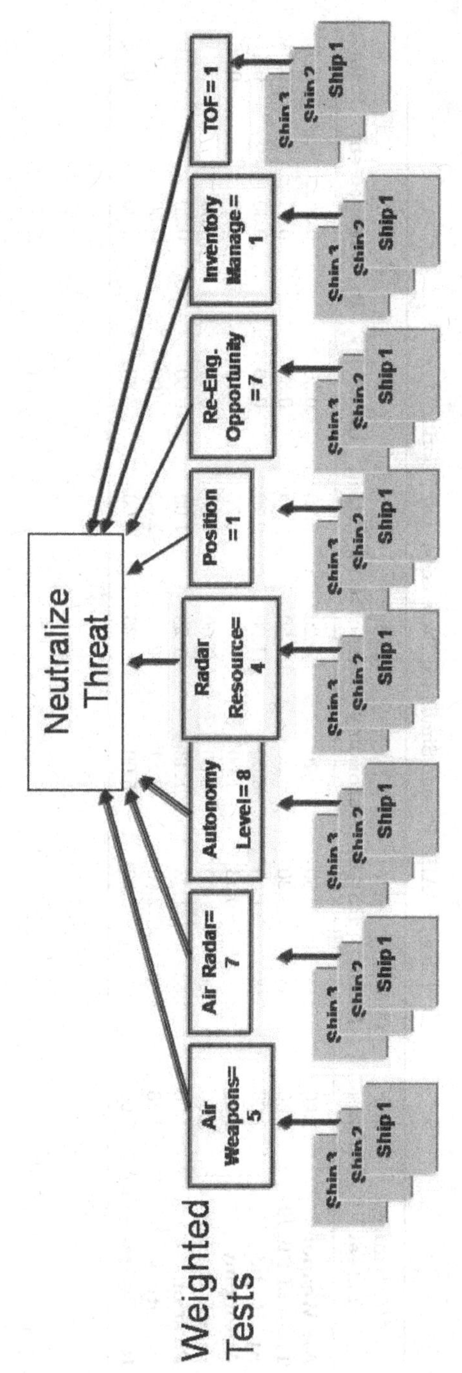

Fig. 5. ABMA Pseudo Simulated Preferences.

Table 5. AHP Pseudo Simulated Preferences and Priorities.

Neutralize Threat	Air Weapons	Time of Flight	Air Radar	Autonomy	Radar Resource	Position	Inventory	Re-Engagement	Priority
Air Weapons	1.00	2.00	0.50	0.70	0.20	0.50	0.30	0.20	0.05
Time of Flight	0.50	1.00	0.20	0.50	0.50	0.30	0.30	0.14	0.04
Air Radai	2.00	5.00	1.00	3.00	1.00	0.70	0.50	0.30	0.11
Autonomy	1.43	2.00	0.33	1.00	0.50	0.70	0.14	0.20	0.06
Radar Resonrce	5.00	2.00	1.00	2.00	1.00	0.80	0.50	0.30	0.11
Position	2.00	3.33	1.43	1.43	1.25	1.00	0.50	0.20	0.10
Inventory	3.33	3.33	2.00	7.14	2.00	2.00	1.00	0.30	0.19
Re-Eiigagement	5.00	7.14	3.33	5.00	3.33	5.00	3.33	1.00	0.35

A Lockheed Martin and a preliminary Stevens Institute of Technology agent based simulation experiment of a battle group encountering several air threats demonstrated the viability of MCDA in a SoS problem domain revealing that the social function adjusts based on available resources. The autonomous systems communicate their preferences continuously in the background (BG) comparing against commander's intent at node A1 in Figure 6. For example, C1 node, a battle group ship, requires autonomous utility relief in which to collaborate and communicates its preference to the B1 node as message (1). The B1 node has preference knowledge of nodes C2 and C3, which are other ships or major resources of the battle group such as radar or weapons inventory and communicates with node A1 as message (2). Resources C2 and C3 cannot collaborate as their strict autonomy is their preference. Node A1 or commander's intent has knowledge of nodes B2 and B3 and sends a message (3) to node B1, which in turn communicates with node C1 on collaboration status with nodes B2 or B3. Node C1 receives collaboration preference feedback as message (4) that partnership is possible with nodes B2 and B3.

A summary of results shown in Figure 7 reveals that battle group collaboration improved as ABMA1 and ABMA2 versus the baseline of status quo for a battle group. The three trials consisted of fifty Monte Carlo runs and shows that the battle group neutralizes air threats well given distance. As threats near the assets of the battle group, more cooperation is required as threats approach battle group assets. In ABMA1, the process shown in Figure 7 executes per the same threat and the same seed in the Monte Carlo simulation runs. However, there is improvement as threats are neutralized earlier and at greater distance upon detection shown in

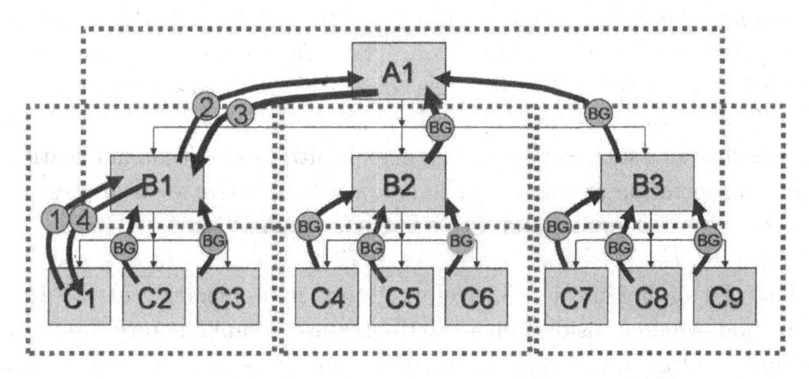

Fig. 6. SoS ABMA MCDA Function.

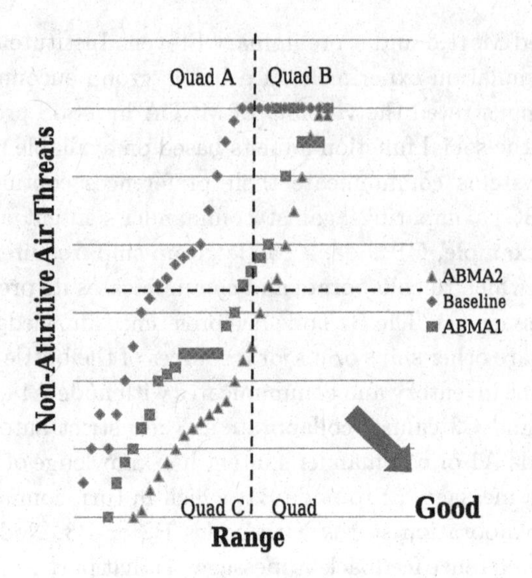

Fig. 7. Battle Group Simulation Results Summary.

Quad B. The ABMA2 scenario is the same as ABMA1 but with air assets. Collaboration is shared between ship based resources and air resources. There is again improvement in neutralizing threats at distance and earlier. The area of Quad D is an ideal goal whereby threats are neutralized early and at greater distance.

Future Research

The SoS systems engineering domain is fairly new and this paper discusses a decision theory approach to the formation of SoS. Further research is required regarding SoS formation. There may be a wide spectrum of research approaches and with many being eclectic given the topic domain of the SoS. Future research in the form of the application of various SoS subject domains such as battle managers, health care, telecommunications, and transportation are needed. The following suggestive areas are research directions to understand and assess SoS social functions:

- The use of nonlinear additive weighting in a hierarchical weighting method would consider when attributes are complimentary enhancing another attribute, when substitutes an attribute, or reduces the performance of other attributes. For simple additive weighting methods, it is assumed that the utility value of the multiple attributes represent the individual attributes [42, 43].

- The Technique for Order Preference by Similarity to Ideal Solution (TOPSIS) is based on the idea that the chosen alternative shall have the shortest distance from the ideal solution and the farthest distance from the negative-ideal solution [43]. This method has been shown to provide preference order solutions and is similar to the Hurwicz rule because it is an amalgamation of optimistic and pessimistic decision methods [44]. This may solve the problem of non-normalized data as is required for the Hurwicz rule. The comparative analysis for a SoS is necessary.
- Fuzzy logic approach to MCDA in the SoS social function is an important approach to determine system preferences for collaboration as the preferences become more complex and humanistic. These systems and their description of interoperability become nebulous and incomparable. Nebulous goals and constraints result in nebulous or fuzzy attributes. There are techniques described in the literature that transform fuzzy attributes into discrete identifiers, but are not well suited for system choices of collaboration [45]. The complexity of the interoperability of systems leads to greater nebulous reasons of belonging and eventual collaboration. Probabilities between 0 and 1 correspond to degrees of belief for a Bayesian approach. However, degrees of truth is different than degrees of belief and may be approached using fuzzy logic [20]. The approach for SoS social utility function may be described from [46] as an approach for preference evaluation.

Conclusion

This paper has discussed at length the concept of a SoS collaborative formation mechanism whereby there exists an *invisible hand,* initially described by Adam Smith, that has an observable and patterned effect on processes and systems. The case study and associated simulation demonstrates a real-time SoS whereby a disparate group of autonomous systems are directed to collaborate while adjudicating their utility functions via their set of preferences. A SoS run-time is revealed in the simulation by autonomous systems asking for collaboration and other systems respond with their ability to collaborate, but mitigated by a SoS goal. In the simulation and case study, there is an important nuance:

1. Autonomous systems collaborate given their preferences and utility function upon a need to do so with their preferences and utility function

satisficed to gain collaboration. They are allowed to collaborate due to a collective social function or one that is predetermined, being commander's intent in the case study, or an immediate need by a single autonomous system.

2. There is an amorphous SoS capability not visibly demonstrated in the simulation, but discussed in the case study. This situation in the ABMA case study becomes present without commander's intent when autonomous systems seek collaboration out of a need for resources or aid to meet their own utility function. The process discussed here affords the capability to collaborate without a SoS Command and Control (C2) and is direction of future research.

References

1. A. Smith, "An Inquiry into the Nature and Causes of the Wealth of Nations," 5 ed, E. Cannan, Ed. London: Methuen & Co., Ltd., 1776.
2. R. J. Cloutier, M. J. DiMario, and H. W. Polzer, "Net-centricity and System of Systems," in *System of Systems: Innovations for the 21st Century*, M. Jamshidi, Ed. Hoboken, NJ: John Wiley & Sons, 2009, pp. 150–168.
3. J. M. Tien, "A System of Systems View of Services," in *System of Systems Engineering: Innovations for the 21st Century*, M. Jamshidi, Ed. Hoboken, NJ: John Wiley & Sons, 2009, pp. 293–316.
4. C. Keating, R. Rogers, R. Unal, D. Dryer, A. Sousa-Poza, R. Safford, W. Peterson, and G. Rabadi, "System of Systems Engineering," *Engineering Management Journal*, vol. 15, pp. 36–45, September 2003.
5. L. v. Bertalanffy, *General System Theory: Foundations, Development, Applications.* New York: George Braziller, 1969.
6. K. Boulding, "General Systems Theory: The Skeleton of Science," *Management Science,* vol. 2, pp. 197–208, April 1956.
7. J. Boardman and B. Sauser, "System of Systems — the meaning of *of*," in *IEEE International Conference on System of Systems Engineering*, Los Angeles, CA, 2006, pp. 118–123.
8. M. W. Maier, "Architecting Principles for Systems-of-Systems," in *Proceeding of the 6th Annual INCOSE Symposium*, 1996, pp. 567–574.
9. A. P. Sage and C. D. Cuppan, "On the Systems Engineering and Management of Systems of Systems and Federation of Systems," *Information Knowledge Systems Management*, vol. 2, pp. 325–345, 2001.
10. D. A. DeLaurentis and W. A. Crossley, "A Taxonomy-based Perspective for Systems of Systems Design Methods," in *IEEE International Conference on Systems, Man and Cybernetics*, 2005, pp. 86–91.
11. D. A. DeLaurentis, J.-H. Lewe, and D. P. Schrage, "Abstraction and Modeling Hypothesis for Future Transportation Architectures," in *AIAA/CAS International Air and Space Symposium and Exposition: The Next 100 Years*, Dayton, Ohio, 2003.

12. U.S. Department of Defense, "Systems Engineering Guide for System of Systems," 1.0 ed, T. Office of the Under Secretary of Defense for Acquisition, and Logistics, Ed. Washington D.C., 2008.
13. M. J. DiMario, R. J. Cloutier, and D. Verma, "Applying Frameworks To Manage System of Systems Architecture," *Engineering Management Journal*, vol. 20, pp. 18–23, December 2008.
14. P. J. Driscoll, "Systems Thinking," in *Decision Making in Systems Engineering and Management*, G. S. Parnell, P. J. Driscoll, and D. L. Henderson, Eds. Hoboken, NJ: John Wiley & Sons, 2008, pp. 19–53.
15. P. D. West, J. E. Kobza, and S. R. Goerger, "Systems Modeling and Analysis," in *Decision Making in Systems Engineering and Management*, G. S. Parnell, P. J. Driscoll, and D. L. Henderson, Eds. Hoboken, NJ: John Wiley & Sons, 2008, pp. 79–118.
16. H. A. Simon, "A Behavioral Model of Rational Choice," *Quarterly Journal of Economics*, vol. 59, pp. 99–118, 1955.
17. A. Gorod, M. J. DiMario, B. Sauser, and J. Boardman, "Satisficing" System of Systems (SoS) using Dynamic and Static Doctrine," *International Journal of System of Systems Engineering (IJSSE)*, Submitted for publication.
18. W. C. Stirling and R. L. Frost, "Social Utility Functions-Part II: Applications," *IEEE Transactions on Systems, Man and Cybernetics, Part C: Applications and Reviews*, vol. 35, pp. 533–543, November 2005.
19. R. B. Myerson, "Mechanism Design," *Econometrica*, pp. 1767–97, 1984.
20. S. J. Russell and P. Norvig, *Artificial Intelligence: A Modern Approach.* Upper Saddle River: Pearson Education, 2003.
21. L. Hurwicz, "The Design of Mechanisms for Resource Allocation," *The American Economic Review*, vol. 63, pp. 1–30, May 1973.
22. J. K. Archibald, J. C. Hill, F. R. Johnson, and W. C. Stirling, "Satisficing Negotiations," *IEEE Transactions on Systems, Man and Cybernetics, Part C: Applications and Reviews*, vol. 36, pp. 4–18, January 2006.
23. J. O. Ledyard, "Incentive Compatibility," in *The New Palgrave: Allocation, Information, and Markets*, J. Eatwell, M. Milgate, and P. Newman, Eds. New York: W. W. Norton, 1989, pp. 141–151.
24. K. H. Kim and F. W. Roush, *Team Theory.* New York, NY: John Wiley & Sons, 1987.
25. J. Boardman and B. Sauser, *Systems Thinking: Coping with 21st Century Problems.* Boca Raton: CRC, 2008.
26. B. Sauser, J. Boardman, and A. Gorod, "System of Systems Management," in *System of Systems: Innovations for the 21st Century*, M. Jamshidi, Ed. Hoboken, NJ: John Wiley & Sons, 2009, pp. 191–217.
27. A. Gorod, B. Sauser, and J. Boardman, "System-of-Systems Engineering Management: A Review of Modern History and a Path Forward," *IEEE Systems Journal*, vol. 2, pp. 484–499, December 2008.
28. W. C. Stirling, "Social Utility Functions-Part I: Theory," *IEEE Transactions on Systems, Man and Cybernetics, Part C: Applications and Reviews,* vol. 35, pp. 522–532, November 2005.

29. W. C. Stirling, *Satisficing Games and Decision Making: With Applications to Engineering and Computer Science*. Cambridge: Cambridge University, 2003.
30. S. Chao, *Creating a Business Case for Quality Improvement Research: Expert Views, Workshop Summary*. Washington D.C.: National Academies, 2008.
31. V. Belton and T. J. Stewart, *Multiple Criteria Decision Analysis: An Integrated Approach*. Boston, MA: Kluwer Academic Publishers, 2002.
32. T. L. Saaty, *The Analytic Hierarchy Process*. New York: McGraw-Hill, 1980.
33. J. W. Lee and S. H. Kim, "Using analytic network process and goal programming for interdependent information system project selection," *Computers and Operational Research*, vol. 27, pp. 367–382, April 2000.
34. M. Klein, H. Sayama, P. Faratin, and Y. Bar-Yam, "The dynamics of collaborative design: Insights from complex systems and negotiation research," *Concurrent Engineering Research and Applications Journal*, vol. 12, 2003.
35. T. L. Saaty, "How to make a decision: The Analytic Hierarchy Process," *European Journal of Operational Research*, vol. 48, pp. 9–26, 1990.
36. U.S. Congress, *SDI: Technology, Survivability, and Software*. Washington D.C.: US Government Printing Office, 1988.
37. U.S. Department of Defense, "Joint Pub 1-02," in *Department of Defense Dictionary of Military and Associated Terms*, 2001.
38. U.S. Joint Chiefs of Staff, "Joint Vision 2020," Government Printing Office, Washington D.C., 2000.
39. U.S. Government Accountability Office, "Defense Acquisitions: Challenges Associated with the Navy's Long-Range Shipbuilding Plan," Washington D.C.: U.S. Government Accountability Office, 2006.
40. U.S. Government Accountability Office, "Military Readiness: Navy's Fleet Response Plan Would Benefit from a Comprehensive Management Approach and Rigorous Testing," Washington D.C.: U.S. Government Accountability Office, 2005.
41. U.S. Department of Defense, "Command and Control Joint Integrating Concept," Washington D.C., 2005.
42. C.-L. Hwang and M.-J. Lin, *Group Decision Making under Multiple Criteria: Methods and Applications*. New York, NY: Springer-Verlag, 1987.
43. C.-L. Hwang and K. Yoon, *Multiple Attribute Decision Making: Methods and Applications, A State-of-the-Art Survey*. New York, NY: Springer-Verlag, 1981.
44. J. D. Hey, *Uncertainty in Microeconomics*. Oxford, England: Martin Robertson, 1979.
45. S.-J. Chen, C.-L. Hwang, and F. P. Hwang, *Fuzzy Multiple Attribute Decision Making: Methods and Applications*. New York, NY: Springer-Verlag, 1992.
46. D. Verma, C. Smith, and W. Fabrycky, "Fuzzy Set Based Multi-Attribute Conceptual Design Evaluation," *Systems Engineering*, vol. 2, pp. 187–197, January 2000.

Appendix H

SYSTEM OF SYSTEMS MULTICRITERIA DECISION AID AUTO BATTLE MANAGEMENT SIMULATION ARCHITECTURE

The simulation performed consisted of four missile ships with self-defense capabilities, human-machine interface capabilities, local autonomy battle management, and composite auto battle management or SoS battle management services. These services allowed for MCDA and collaborative schemes. MCDA ship preferences and ship resource status was shared with local autonomy battle managers.

A Service-Oriented Architecture (SOA) and High Level Architecture (HLA) were used to interconnect the ship simulations and sensor simulations with fusion engines to process data. The processed data was shared with the four ship simulations and air assets represented in Figure 35.

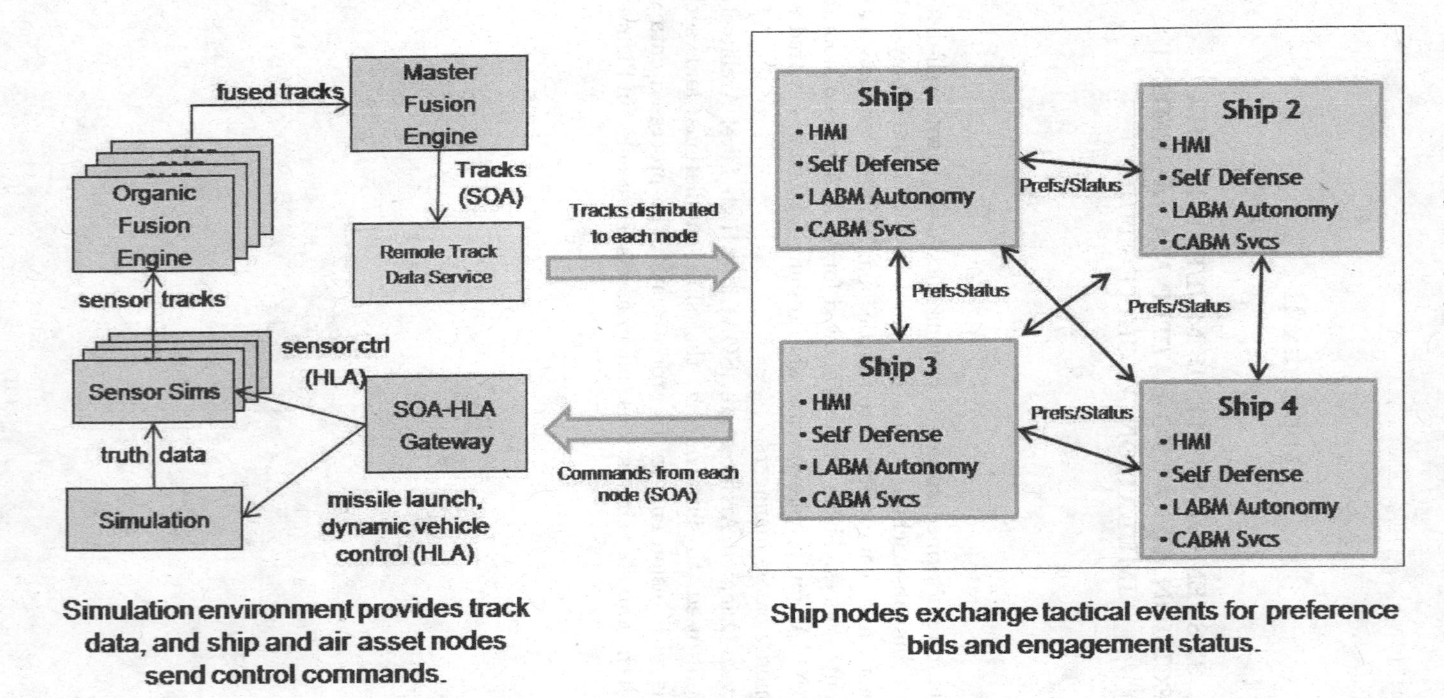

Fig. 35. SoS ABMA Simulation Environment Architecture.

REFERENCES

1. Acquisition Solutions, I. (2001). *Independent Assessment of the United States Coast Guard "Integrated Deepwater System" Acquisition*: Defense Acquisition University.
2. Alberts, D. S., Garstka, J. J. & Stein, F. P. (2000). *Network Centric Warfare: Developing and Leveraging Information Superiority*. Retrieved. from http://dodccrp.org/html4/books_downloads.html.
3. Alberts, D. S. & Hayes, R. E. (2003). *Power to the Edge: Command... Control... in the Information Age*. Retrieved August 12, 2006. from http://dodccrp.org/html4/books_downloads.html.
4. Aldridge, J. E. C. (2004). *Report of the President's Commission on Implementation of United States Space Exploration Policy: A Journey to Inspire, Innovate, and Discover*. Retrieved April 8, 2009. from http://www.nasa.gov/pdf/60736main_M2M_report_small.pdf.
5. Archibald, J. K., Hill, J. C., Johnson, F. R. & Stirling, W. C. (2006). Satisficing Negotiations. *IEEE Transactions on Systems, Man and Cybernetics, Part C: Applications and Reviews*, 36(1), 4–18.
6. Arnberg, M. (2005). Integrated Deepwater Systems (IDS): SoS Challenges in USCG Deepwater Program.
7. Arquilla, J. & Ronfeldt, D. (2001). The Advent Of Netwar (Revisited). In J. Arquilla & D. Ronfeldt (Eds.), *Networks and Netwars: The Future of Terror, Crime, and Militancy* (pp. 1–25). Santa Monica: Rand.
8. Arrow, H., McGrath, J. E. & Berdahl, J. L. (2000). *Small Groups As Complex Systems: Formation, Coordination, Development, and Adaptation*. Thousand Oaks: Sage Publications.
9. Arrow, K. J. (1977). Decentralization Within Firms. In K. J. Arrow & L. Hurwicz (Eds.), *Studies in resource allocation processes* (pp. 134–145). New York, NY: Cambridge University Press.
10. Axelrod, R. (2006). *The Evolution Of Cooperation*. New York: Basic Books.
11. Barabasi, A.-L. (2003). *Linked*. London: Penguin Group.
12. Baran, P. (1964). On Distributed Communications: 1. Introduction To Distributed Communications Networks, pp. 37). Available from http://www.rand.org/pubs/research_memoranda/RM3420/RM3420.chapter1.html
13. Bassler, B. L. & Losick, R. (2006). Bacterially Speaking. *Cell*, 125, 237–246.
14. Belton, V. & Stewart, T. J. (2002). *Multiple Criteria Decision Analysis: An Integrated Approach*. Boston, MA: Kluwer Academic Publishers.
15. Bertalanffy, L. v. (1969). *General System Theory: Foundations, Development, Applications*. New York: George Braziller.

16. Boardman, J. (1995). Wholes and Parts — A Systems Approach. *IEEE Transactions on Systems, Man and Cybernetics*, 25(7), 1150–1161.
17. Boardman, J. & Sauser, B. (2006, April 2006). *System of Systems — the meaning of of.* Paper presented at the IEEE International Conference on System of Systems Engineering, Los Angeles, CA.
18. Boardman, J. & Sauser, B. (2008). *Systems Thinking: Coping with 21st Century Problems.* Boca Raton: CRC.
19. Boardman, J., Sauser, B., John, L. & Edson, R. (2008). *The Conceptagon: A Framework for Systems Thinking and Systems Practice.* Paper presented at the IEEE International Conference on Systems, Man, and Cybernetics.
20. Bonabeau, E. & Meyer, C. (2001). Swarm Intelligence: A Whole New Way to Think About Business. *Harvard Business Review.*
21. Boulding, K. (1956). General Systems Theory: The Skeleton of Science. *Management Science*, 2(3), 197–208.
22. Buchanan, J. T., Henig, E. J. & Henig, M. I. (1998). Objectivity and subjectivity in the decision making process. *Annals of Operations Research*, 80, 333–345.
23. Buede, D. M. (2000). *The Engineering Design of Systems: Models and methods.* New York: John Wiley & Sons, Inc.
24. Caffall, D. S. (2005). *Developing Dependable Software for a System-of-Systems.* Unpublished PhD, Naval Postgraduate School, Monterey.
25. Carney, D., Fisher, D., Morris, E. & Place, P. (2005). *Some Current Approaches to Interoperability* (No. CMU/SEI-2005-TN-033). Pittsburgh: Software Engineering Institute.
26. Carney, D., Fisher, D. & Place, P. (2005). *Topics in Interoperability: System-of-Systems Evolution.* Pittsburgh: Software Engineering Institute.
27. Chao, S. (2008). *Creating a Business Case for Quality Improvement Research: Expert Views, Workshop Summary.* Washington D.C.: National Academies.
28. Checkland, P. (1993). *Systems Thinking, Systems Practice.* New York: John Wiley & Sons, Inc.
29. Checkland, P. (2000). Soft Systems Methodology: A Thirty Year Retrospectivie. *Systems Research and Behaviorial Science*, 17, S11–S58.
30. Chen, P. & Clothier, J. (2003). Advancing Systems Engineering for Systems-of-Systems Challenges. *Systems Engineering*, 6(3), 170–183.
31. Chen, S.-J., Hwang, C.-L. & Hwang, F. P. (1992). *Fuzzy Multiple Attribute Decision Making: Methods and Applications.* New York, NY: Springer-Verlag.
32. Christensen, C. M., Grossman, J. H. & Hwang, J. (2009). *The Innovator's Prescription: A Disruptive Solution for Health Care.* New York, NY: McGraw-Hill.
33. Cloutier, R. J., DiMario, M. J. & Polzer, H. W. (2009). Net-centricity and System of Systems. In M. Jamshidi (Ed.), *System of Systems: Innovations for the 21st Century* (pp. 150–168). Hoboken, NJ: John Wiley & Sons.
34. Corner, J., Buchanan, J. & Henig, M. (2001). Dynamic Decision Problem Structuring. *Journal of Multi-Criteria Decision Analysis*, 10, 129–141.

35. Cortes, F., Przeworski, A. & Sprague, J. (1974). *Systems Analysis for Social Scientists*. New York: John Wiley & Sons.
36. Crutchfiled, J. P., Mitchell, M. & Das, R. (2003). The Evolutionary Design of Collective Computation in Cellular Automata. In J. P. C. a. P. K. Schuster (Ed.), *Evolutionary Dynamics: Exploring the Interplay of Selection, Neutrality, Accident, and Function* (pp. 361–411). New York: Oxford University Press.
37. De Wolf, T. (2007). *Analysing and Engineering Self-Organizing Emergent Applications*. Unpublished Doctorate, Katholieke Universiteit Leuven, Leuven, Belgium.
38. Degenne, A. & Forse, M. (1999). *Introducing Social Networks* (A. Borges, Trans.). London: Sage Publications.
39. DeLaurentis, D. A. & Crossley, W. A. (2005, 10–12 October 2005). *A Taxonomy-based Perspective for Systems of Systems Design Methods*. Paper presented at the IEEE International Conference on Systems, Man and Cybernetics.
40. DeLaurentis, D. A., Dickerson, C., DiMario, M. J., Gartz, P., Jamshidi, M. M., Nahavandi, S., *et al.* (2007). A Case for an International Consortium on System-of-Systems Engineering. *IEEE Systems Journal*, 1(1), 68–73.
41. DeLaurentis, D. A., Fry, D. N., Oleg, V. S. & Ayyalasomayajula, S. (2006). Modeling Framework and Lexicon for System-of-Systems Problems.
42. DeLaurentis, D. A., Lewe, J.-H. & Schrage, D. P. (2003, July 14–17). *Abstraction And Modeling Hypothesis For Future Transportation Architectures*. Paper presented at the AIAA/CAS International Air and Space Symposium and Exposition: The Next 100 Years, Dayton, Ohio.
43. Department of the Navy. (2004). *Modeling and Simulation Verification, Validation, and Accreditation Implementation Handbook*. Retrieved January 20, 2009. from https://nmso.navy.mil/DesktopModules/XSDocumentLibrary/Components/FileDownloader/XSFileDownloaderPage.aspx?tabid=96&xsdid=633&xspid=0&xslrf=/DesktopModules/XSDocumentLibrary/App_LocalResources/XSDocumentLibrary&xscl=en-US&xsmcs=%2FDesktopModules%2FXSDocumentLibrary%2F&xsuid=1&xsufn=Site+Manager&xsur=&xsuarn=Administrators&xscd=False&xstmid=353.
44. DiMario, M. J. (2006, April). *System of Systems Interoperability Types and Characteristics in Joint Command and Control*. Paper presented at the IEEE Conference on System of Systems Engineering, Los Angeles, CA.
45. DiMario, M. J., Cloutier, R. J. & Verma, D. (2008). Applying Frameworks To Manage System of Systems Architecture. *Engineering Management Journal*, 20(4), 18–23.
46. DoD. (2006). *System of Systems Systems Engineering Guide: Considerations for Systems Engineering in a System of Systems Environment*. Retrieved October 27, 2007. from https://acc.dau.mil/CommunityBrowser.aspx?id=147494&cid=17608&lang=en-US.
47. Driscoll, P. J. (2008). Systems Thinking. In G. S. Parnell, P. J. Driscoll & D. L. Henderson (Eds.), *Decision Making in Systems Engineering and Management* (pp. 19–53). Hoboken, NJ: John Wiley & Sons.

48. Figueira, J., Salvatore, G. & Ehrgott, M. (2005). *Multiple Criteria Decision Analysis: State Of The Art Surveys.* New York, NY: Springer Science+Business Media, Inc.
49. Fisher, D. A. (2006). *An Emergent Perspective on Interoperation in Systems of Systems* (No. CMU/SEI-2006-TR-003). Pittsburgh: Carnegie Mellon Software Engineering Institute.
50. Flood, R. L. (1999). *Rethinking The Fifth Discipline: Learning Within the Unknowable* (2002 ed.). London: Routledge.
51. Ganek, A. G. & Corbi, T. A. (2003). The Dawning of the Autonomic Computing Era. *IBM Systems Journal*, 42(1), 5–18.
52. Garnier, S., Gautrais, J. & Theraulaz, G. (2007). The Biological Principles of Swarm Intelligence. *Swarm Intelligence*, 1(1), 3–31.
53. Goodrich, M. A., Stirling, W. C. & Frost, R. L. (1998). A Theory of Satisficing Decisions and Control. *IEEE Transactions on Systems, Man and Cybernetics-Part A: Systems and Humans*, 28(6), 763–779.
54. Gorod, A., DiMario, M. J., Sauser, B. & Boardman, J. "Satisficing" System of Systems (SoS) using Dynamic and Static Doctrine. *International Journal of System of Systems Engineering (IJSSE)*, 1(3), 347–366.
55. Gorod, A., Sauser, B. & Boardman, J. (2008). System-of-Systems Engineering Management: A Review of Modern History and a Path Forward. *IEEE Systems Journal*, 2(4), 484–499.
56. Graziano, A. M. & Raulin, M. L. (2004). *Research Methods: A Process of Inquiry* (5 ed.). Boston: Pearson Education Group.
57. Groves, T. (1973). Incentives in Teams. *Econometrica*, 41(4), 617–631.
58. Groves, T. & Ledyard, J. O. (1977). Some Limitations of Demand Revealing Processes. [Discussion Paper No. 219]. *Public Choice*, 29(2), 107–124.
59. Hey, J. D. (1979). *Uncertainty in Microeconomics.* Oxford, England: Martin Robertson.
60. Hobbs, B. F. & Meier, P. (2000). *Energy Decisions and the Environment: A Guide to the Use of Multicriteria Methods.* Boston, MA: Kluwer Academic Publishers.
61. Holland, J. H. (1995). *Hidden Order: How Adaptation Builds Complexity.* New York: Basic Books.
62. Holland, J. H. (1998). *Emergence: From Chaos to Order.* New York: Basic Books.
63. Huang, H.-M., Pavek, K., Albus, J. & Messina, E. (2005). *Autonomy Levels for Unmanned Systems (ALFUS) Framework: An Update.* Paper presented at the 2005 SPIE Defense and Security Symposium. Retrieved November 18, 2006, from http://www.isd.mel.nist.gov/projects/autonomy_levels/.
64. Hurwicz, L. (1973). The Design of Mechanisms for Resource Allocation. *The American Economic Review*, 63(2), 1–30.
65. Hurwicz, L. & Reiter, S. (2006). *Designing Economic Mechanisms.* Cambridge: Cambridge University Press.
66. Hwang, C.-L. & Lin, M.-J. (1987). *Group Decision Making under Multiple Criteria: Methods and Applications.* New York, NY: Springer-Verlag.

67. Hwang, C.-L. & Yoon, K. (1981). *Multiple Attribute Decision Making: Methods and Applications, A State-of-the-Art Survey.* New York, NY: Springer-Verlag.
68. INCOSE. (2007). *INCOSE Systems Engineering Handbook V.3.1* (3.1 ed.): INCOSE.
69. Institute of Medicine. (2001). *Crossing The Quality Chasm: A New Health System for the 21st Century* (2 ed.). Washington DC: National Academy Press.
70. ISO. (2002). ISO/IES 15288 Systems Engineering — System Life Cycle Processes (1 ed.). Switzerland: ISO/IEC.
71. Jensen, F. V. & Nielsen, T. D. (2007). *Bayesian Networks and Decision Graphs* (2 ed.). New York, NY: Springer.
72. Johnson, S. (2001). *Emergence.* New York: Scribner.
73. Karni, R., Sanchez, P. & Tummala, V. M. R. (1990). A Comparative Study of Multiattribute Decision Making Methodologies. *Theory and Decision*, 29, 203–222.
74. Keating, C., Rogers, R., Unal, R., Dryer, D., Sousa-Poza, A., Safford, R., *et al.* (2003). System of Systems Engineering. *Engineering Management Journal*, 15(3), 36–45.
75. Keeney, R. L. (1992). *Value-Focused Thinking: A Path to Creative Decision making.* Cambridge, MA: Harvard University Press.
76. Keeney, R. L. & Raiffa, H. (1993). *Decisions with Multiple Objectives: Preferences and Value Tradeoffs.* Cambridge, UK: Cambridge University Press.
77. Kim, K. H. & Roush, F. W. (1987). *Team Theory.* New York, NY: John Wiley & Sons.
78. Koestler, A. (1967). *The Ghost In The Machine* (1 ed.). New York: The Macmillan Company.
79. Krishna, V. & Perry, M. (1998). Efficient Mechanism Design. University Park, PA: Pennsylvania State University.
80. Krygiel, A. J. (1999). *Behind the Wizard's Curtain: An Integration Environment for a System of Systems.* Retrieved June 3, 2006. from http://www.dodccrp.org/publications/pdf/Krygiel_Wizards.pdf.
81. Ledyard, J. O. (1989). Incentive Compatibility. In J. Eatwell, M. Milgate & P. Newman (Eds.), *The New Palgrave: Allocation, Information, and Markets* (pp. 141–151). New York: W. W. Norton.
82. Lee, J. W. & Kim, S. H. (2000). Using Analytic Network Process and Goal Programming for Interdependent Information System Project Selection. *Computers and Operational Research*, 27(4), 367–382.
83. Maier, M. W. (1996). *Architecting Principles for Systems-of-Systems.* Paper presented at the Proceeding of the 6th Annual INCOSE Symposium.
84. Majumdar, W. S. (2007). *System of Systems Technology Readiness Assessment.* Unpublished Masters, Naval Postgraduate School, Monterey.
85. Metron. (2008). Industry Support: DDG-1000 Design and Analyses. Retrieved March 12, 2009, from http://www.metsci.com/Default.aspx?tabid=201

86. Midgley, G. (2006). Systemic Intervention for Public Health. *American Journal of Public Health*, 96(3), 466–472.
87. Mitchell, M. (1998). *An Introduction to Genetic Algorithms.* Cambridge: MIT.
88. Mitchell, M. (2001). Life and Evolution in Computers. *History and Philosophy of the Life Sciences*, 23, 361–383.
89. MODAF Development Team. (2008). *The MOD Architecture Framework Version 1.2.* Retrieved March 2, 2009. from http://www.modaf.org.uk/.
90. Morganwalp, J. (2002). *A System of Systems Focused Enterprise Architectural Framework.* Unpublished PHD, George Mason University, Fairfax, VA.
91. Morris, E., Levine, L., Meyers, C., Place, P. & Plakosh, D. (2004). *System of Systems Interoperability (SOSI): Final Report* (No. CMU/SEI-2004-TR-004). Pittsburgh: Software Engineering Institute.
92. Myerson, R. B. (1984). Mechanism Design. *Econometrica* (51), 1767–1797.
93. Nash, J. (1951). Non-Cooperative Games. *The Annals of Mathematics*, 54(2), 286–295.
94. Nash, J. F. (1950). The Bargaining Problem. *Econometrica*, 18(2), 155–162.
95. Owens, W. A. (1996). *The Emerging U.S. System-of-Systems* (No. 63). Washington D.C.: National Defense University Institute For National Strategic Studies.
96. Parunak, H. V. D. (1997). "Go to the Ant": Engineering Principles from Natural Multi-Agent Systems. *Annals of Operations Research*, 75, 69–101.
97. Porter, M. E. & Teisberg, E. O. (2006). *Redefining Health Care: Creating Value-Based Competition on Results.* Boston: Harvard Business School Press.
98. Reid, P. P., Compton, W. D., Grossman, J. H. & Fanjiang, G. (2005). *Building A Better Delivery System: A New Engineering/Health Care Partnership.* Washington D.C.: The National Academies Press.
99. Russell, S. J. & Norvig, P. (2003). *Artificial Intelligence: A Modern Approach.* Upper Saddle River: Pearson Education.
100. Saaty, T. L. (1980). *The Analytic Hierarchy Process.* New York: McGraw-Hill.
101. Saaty, T. L. (2005). *Theory and Applications of the Analytical Network Process: Decision Making with Benefits, Opportunities, Costs, and Risks.* Pittsburgh, PA: RWS Publications.
102. Saaty, T. L. (2008). *Decision Making For Leaders.* Pittsburgh, PA: RWS Publications.
103. Saaty, T. L. & Vargas, L. G. (1991). *Prediction, Projection And Forecasting: Applications of the Analytic Hierarchy Process in Economics, Finance, Politics, Games, and Sports.* Boston: Kluwer Academic Publishers.
104. Saaty, T. L. & Vargas, L. G. (2001). *Models, Methods, Concepts & Applications of the Analytic Hierarchy Process.* Boston: Kluwer Academic Publishers.

105. Sage, A. P. & Cuppan, C. D. (2001). On the Systems Engineering and Management of Systems of Systems and Federation of Systems. *Information Knowledge Systems Management*, 2(4), 325–345.
106. Saunders, T., Croom, C., Austin, W., Brock, J., Crawford, N., Endsley, M., *et al.* (2005). *System-of-Systems Engineering for Air Force Capability Development: Executive Summary and Annotated Brief.* Retrieved April 3, 2007. from http://oai.dtic.mil/oai/oai?verb=getRecord&metadataPrefix=html&identifier=ADA442612.
107. Senge, P. M., Kleiner, A., Roberts, C., Ross, R. B. & Smith, B. J. (1994). *The Fifth Discipline Fieldbook: Strategies and Tools For Building A Learning Organization.* New York: Doubleday.
108. Simon, H. A. (1955). A Behavioral Model of Rational Choice. *Quarterly Journal of Economics*, 59(1), 99–118.
109. Simon, H. A. (1956). Rational Choice And The Structure of the Environment. *Psychological Review*, 63(2), 129–138.
110. Simon, H. A. (1962). The Architecture of Complexity. *Proceedings American Philosophical Society*, 106(6), 467–482.
111. Simon, H. A. (1996). *The Sciences of the Artificial* (3 ed.). Cambridge: The MIT Press.
112. Smith, A. (1776). An Inquiry into the Nature and Causes of the Wealth of Nations. In E. Cannan (Eds.), Retrieved from Library of Economics and Liberty database Available from http://www.econlib.org/library/Smith/smWN13.html.
113. Starr, P. (1982). *The Social Transformation of American Medicine*: Basic Books.
114. Sterman, J. D. (2006). Learning from Evidence in a Complex World. *American Journal of Public Health*, 96(3), 505–414.
115. Stirling, W. C. (2003). *Satisficing Games and Decision Making: With Applications to Engineering and Computer Science.* Cambridge: Cambridge University.
116. Stirling, W. C. (2005). Social Utility Functions-Part I: Theory. *IEEE Transactions on Systems, Man and Cybernetics, Part C: Applications and Reviews*, 35(4), 522–532.
117. Stirling, W. C. & Frost, R. L. (2005). Social Utility Functions — Part II: Applications. *IEEE Transactions on Systems, Man and Cybernetics, Part C: Applications and Reviews*, 35(4), 533–543.
118. Sund, C. (2007). Digital Close Air Support. *Marine Corps Gazette*, 91, 21–28.
119. Tien, J. M. (2009). A System of Systems View of Services. In M. Jamshidi (Ed.), *System of Systems Engineering: Innovations for the 21st Century* (pp. 293–316). Hoboken, NJ: John Wiley & Sons.
120. Truszkowski, W. F., Hinchey, M. G., Rash, J. L. & Rouff, C. A. (2006). Autonomous and Autonomic Systems: A Paradigm for Future Space Exploration Missions. *IEEE Transactions on Systems, Man and Cybernetics — Part C: Applications and Reviews*, 36(3), 279–291.
121. U.S. Department of Defense. (2001). *Joint Pub 1-02.* Retrieved. from www.dtic.mil/doctrine/jel/new_pubs/jp1_02.pdf.

122. U.S. Department of Defense. (2008). *Systems Engineering Guide for System of Systems.* Retrieved June 26, 2008. from www.acq.osd.mil/sse/docs/SE-Guide-for-SoS.pdf.
123. U.S. Joint Chiefs of Staff. (2005). *Joint Capabilities Integration and Development System: CJCSI 3170.01E.* Retrieved July 4, 2006. from http://www.dtic.mil/cjcs_directives/cdata/unlimit/3170_01.pdf.
124. van Diggelen, J. (2007). *Achieving Semantic Interoperability in Multi-agent Systems.* Unpublished PhD, Dutch Research School for Information and Knowledge Systems.
125. Verma, D., Smith, C. & Fabrycky, W. (2000). Fuzzy Set Based Multi-Attribute Conceptual Design Evaluation. *Systems Engineering*, 2(4), 187–197.
126. Von Neumann, J. & Morgenstern, O. (1953). *The Theory of Games and Economic Behavior* (3 ed.). Princeton, NJ: Princeton University Press.
127. Weinberg, G. M. (2001). *An Introduction to General Systems Thinking* (Silver Anniversary ed.). New York: Dorset House Publishing.
128. West, P. D., Kobza, J. E. & Goerger, S. R. (2008). Systems Modeling and Analysis. In G. S. Parnell, P. J. Driscoll & D. L. Henderson (Eds.), *Decision Making in Systems Engineering and Management* (pp. 79–118). Hoboken, NJ: John Wiley & Sons.
129. Williams, A. B. (1999). *Learning Ontologies in a Multiagent System.* Unpublished PHD, Univeristy of Kansas.
130. Winans, S. C. & Bassler, B. L. (2002). Mob Psychology. *Journal of Bacteriology*, 184(4), 873–883.
131. Yoon, K. P. & Hwang, C.-L. (1995). *Multiple Attribute Decision Making: An Introduction.* Thousand Oaks, CA: Sage Publications.